Central Heating

Installation, Maintenance and Repair

First published in Great Britain 2008 *by Brailsford with WritersPrintShop*

ISBN 190462362X

Cover illustration: open vented y plan system showing the hot water cylinder, motorised valve, pump and vent chamber.

Designed by e-BookServices.com

Central Heating

Installation, Maintenance and Repair

Patrick Mitchell

Illustrations and diagrams

1.1 The Hypocaust system at Housteads fort on Hadrian's Wall 4
1.2 Cutaway diagram of a modern masonry heater 4
1.3 Illustration of a boiler made up of cast-iron sections 6
2.1 Single pipe central heating system 14
2.2 Two pipe pumped central heating system 14
2.3 The effects of pumping water on pressures in an open vented system 18
2.4 Graph showing how properties of water change with temperature 19
2.5 Radiator connection in a single pipe system 20
2.6 The 3T layout of open vented central heating systems 30
2.7 The 4T 2T layout of open vented central heating systems 31
2.8 A variant on the 4T 2T layout 31
2.9 Separate connections for the hot water and space heating 32
2.10 A vent chamber of trap 34
2.11 Expansion chamber used in sealed central heating systems 34
3.1 Conventional gravity fed hot water systems 38
3.2 An arrangement that can be useful in tall houses 41
3.3 A primatic or single feed hot water cylinder 42
3.4 Electric hot water systems 43
3.5 One form of thermal store hot water system 49
3.6 Secondary heat exchanger 49
5.1 Types of pressed steel radiator panel 61
5.2 A cast iron radiator 62
5.3 An unusual early radiator 63
5.4 Thermostatic radiator valve and lock shield 64
5.5 Block diagram of the structure of an under-floor heating system 70
5.6 Wet under-floor heating system 70
5.7 The central control point for the under-floor system 71
6.1 An electrochemical cell with iron and copper electrodes. 74
7.1 Cut-away diagram of a tube bending machine 87
7.2 A small pipe cutter 88
7.3 The business end of a copper pipe deburring tool 90
7.4 Pre formed fittings for copper pipes 90
7.5 Wire brush 91
9.1 Three port mid position valve 105
9.2 With the motor removed 107
9.3 The motor and switches of the V4073 107
9.4 Operational parts of the valve 108

9.5 The manual lever of the V4073 108
10.1 Exploded view at a Peglar Terrier® pump 112
10.2 Pressure – flow diagram for the pump 112
10.3 View of the motor end of the pump 117
11.1 Diagram of flueing arrangements for boilers 120
12.1 Diagram of a cold water storage cistern 131
13.1 Diagram of a room thermostat with an accelerator resistor 138
13.2 Diagram of a hot water cylinder thermostat 140
14.1 Key to the symbols used in the circuit diagrams 147
14.2 Central heating control circuit 148
14.3 Obsolete circuit but many remain in service 149
14.4 With a tank thermostat added 149
14.5 Control circuit for partly pumped central heating systems 150
14.6 Wiring diagram for a W plan circuit 151
14.7 Wiring diagram for a Y plan circuit 152
14.8 Wiring diagram for an S plan circuit with frost protection 153
15.1 An electric shower with a 45 Amp double pole switch 161
16.1 The pilot light safety system 164
16.2 Fault tracing version of diagram 14.6 165
16.3 Fault tracing version of diagram 14.7 166
16.4 Fault tracing version of diagram 14.8 167

Table of Contents

Introduction 1
Chapter 1. History of Central Heating 3
Chapter 2. Heating Circuits 13
Chapter 3. Hot Water Circuits 37
Chapter 4. Health and Safety 53
Chapter 5. Radiators 59
Chapter 6. Corrosion, Lime Scale and Chemical Additives 73
Chapter 7. Plumbing 85
Chapter 8. Heat 97
Chapter 9. Motorised Valves 105
Chapter 10. Pumps 111
Chapter 11. Boilers 119
Chapter 12. Cisterns 129
Chapter 13. Control Systems 137
Chapter 14. Wiring 145
Chapter 15. Other Heating Technologies 157
Chapter 16. Fault Diagnosis 163
Index. 183

Acknowledgement

I am indebted to Chas Jones for his careful and meticulous review of the manuscript and for numerous invaluable suggestions.

Introduction

This book is aimed at those who choose, use, and work on central heating systems be they householders, landlords, architects, DIY enthusiasts, or professional plumbers. The focus is on explanation of the principals involved in the design and operation of systems. Considerable attention is paid to common failure modes and the often difficult area of fault diagnosis.

Central heating systems lack the appeal of solid fuel fires and stoves. They are seen as being purely functional and utilitarian. Changes to a system will generally not be made while it is functioning satisfactorily. With regular servicing they tend to be fairly reliable and these factors combine to make central heating one of the most enduring forms of domestic technology. Examples of systems dating from any period from the 1950s onward remain in use. Improvements in central heating technology have been in the details of boiler design and efficiency, and system control. The basic principle of using water heated in one place to carry heat via a series of suitable pipes around a building has endured for the past 100 years and is likely to remain with us for the foreseeable future. The main expense and work in installing a central heating system is in building this hot water circuit rather than in purchasing the heating equipment. Most systems can expect to see several changes of boiler and control system in their lifetime. Extra attention to quality and detail in this area will pay dividends in the long run.

The type of system that would be installed in a new dwelling nowadays is markedly different from one still in service dating from the 1970s or earlier. This range of technologies makes it important to have an understanding of not only the functioning of contemporary systems but also of their forebears. Moreover earlier systems tended to require a greater degree of knowledge to install and maintain than more modern designs. As a consequence much of the discussion in this book is dedicated to systems that are now either obsolete or becoming so. It is possible markedly to improve the convenience and efficiency of existing installations by relatively simple and inexpensive upgrades. These involve the replacement of simple T connectors with motorised valves and improvement in control systems.

Legislation surrounding installation of central heating systems has seen rapid change in recent years as a consequence of the emergence of global warming as a political issue. Further change in this direction is anticipated. The need for increasing energy efficiency has meant that all newly installed gas-fired boilers in the UK must now be condensing in design. This will apply to oil fired boilers as well from 2007. The main anticipated development in the coming years is Micro Combined Heat and Power or micro CHP. CHP equipment is currently in use in large commercial buildings and improves efficiency by using heat that would otherwise be wasted to generate electricity that is either used in the building or fed back into the National Grid. Economic systems that are designed on the smaller scale of dwellings are currently under development.

CHAPTER 1
History of Central Heating

Central heating systems burn fuel in one place and distribute the heat around a number of other places. The earliest known reference fitting this description is to a system of under-floor passages through which warmed air was driven to heat the Ancient Greek temple of Ephesus in 350 BC. Whether the system ever existed and if so how well it worked is unknown. The first known practical system in the Western world was the Roman Hypocaust. It consisted of a fireplace built at basement level at one end of the building. The hot gas and smoke coming from the fire were directed through a series of channels or an open space under the stonework of the floor (figure 1.1). The stonework was heated by the passing gases and in turn heated the living space above. Several rooms could be heated from a single fire. We do not have exact details of the technical performance of the hypocaust but a closely analogous system is used by modern masonry heaters (figure 1.2). These are solid fuel fired storage heaters. They are convenient to use compared with open fires because they only need to be fired once or twice a day. Heat is stored in the masonry and slowly given off into the room between firings. Masonry heaters have the virtues of using renewable fuel (wood), burning it cleanly, and with a high thermal efficiency (70 – 85%). It is likely that these features were shared by the Roman Hypocaust which would have made it a good solution to the problem of heating a building, even by today's standards.

Though the hypocaust was the first central heating system, the modern systems we are familiar with were not developed from it. It went out of use in the fourth and fifth centuries AD after the withdrawal of the Roman Empire from Northern Europe. Heating technology then reverted to primitive fireplaces for over 1000 years. It was the technological progress made during and after the 17th-century that led directly to the development of modern central heating systems rather than any ideas derived from earlier civilisations.[1]

[1] Further details of the hypocaust and the fireplaces that followed can be found in "Fireplaces: design and construction of domestic open fires" by Patrick Mitchell.

Figure 1.1: The Hypocaust system at Housteads roman fort on Hadrian's Wall in Northumberland. The figure shows the structure which lay beneath what was once the floor of a building. Some of the flagstones of the floor can still be seen. The stone pillars underneath are spaced about 500mm apart. The cavities between them formed the path for hot gasses to flow from the fire to several chimneys that can be seen as slots in the far wall.

Figure 1.2: Cutaway diagram of a modern masonry heater. At the left-hand end is the firebox. The smoke passes through the system of baffles and up the chimney at the right-hand end. A fire is burned in the firebox once or twice a day and heats up the brickwork. This slowly gives up its heat the room. This modern equivalent of the Hypocaust system is a clean, efficient and convenient form of heating.

In the modern era, early experiments with central heating involved steam based systems. This is partly because they can be very simple, involving only a sealed container of water that is heated and connected to pipes conveying the steam to "radiators". A single pipe can be taken from the "boiler" and connected to a series of radiators. Provision must be made for condensed water to drain back to the boiler but in the early experimental systems water trickled back to the boiler through the same pipe in which steam travelled away from the boiler towards the radiator. Another important contributing factor to the early use of steam based systems was the widespread availability of steam generating boilers for use with steam engines or industrial processes.

Around 1594 Hugh Platt suggested a steam based system for heating a greenhouse. This early reference is exceptional and further experimentation does not seem to have been documented until the mid 18th-century when Colonel Coke proposed a system of steam pipes fed from a central boiler to heat the rooms in a house. James Watt, after whom the Watt[2] is named, used a steam based system to heat his study and employed the method to heat a Manchester factory. In 1795 his associate, the Birmingham entrepreneur Matthew Bolton, put a steam based heating system into a friend's house.

Simple steam based systems suffer from a number of drawbacks and they have never been popular for domestic heating in the British Isles. They operate at or above 100°C and this is hot enough to cause significant burns, making the system potentially unsafe. High-pressure steam based systems run hot enough to burn dust leading to an unpleasant smell when they are switched on.

Steam heating systems are not without their advantages. They do not contain large volumes of water so have a low heat capacity and rapid response times. The higher temperature of radiators allows them to be considerably smaller but they must be shielded to prevent people being burned. The system continues to be developed and is quite popular for heating large buildings. It is commonly used for domestic heating in North America.

The system of using hot water flowing around a series of pipes and radiators to carry heat from a boiler to several rooms is a comparatively recent development. Installations began to appear around the turn of the 20th century. Early systems were limited by the lack of readily available, inexpensive and reliable pumps. They used convection to drive water around the circuit. As convective pressure differences are quite small,

[2] The Watt or W is a unit of power. Power is the rate at which energy is used. One W equals one Joule per second.

pipes had to be of wide diameter to allow adequate flow. This meant the whole circuit contained a large volume of water and consequently had a high heat capacity and took a long time to warm up. Early systems also required considerable expert attention to operate. They were mainly installed in large public or commercial buildings rather than domestic houses. This remained largely unchanged until inexpensive electric centrifugal water pumps became widely available in the 1950s. Once water could be pumped round the radiator circuit the performance and installation costs of systems improved markedly. As a consequence, central heating gained popularity in the domestic market.

The first central heating systems were coal-fired, and 200 years later, coal-fired systems are still available. They were commonly used during the first half of the 20th century for gravity fed systems installed in large public or industrial buildings. An example of a boiler dating from that era is shown in figure 1.3. In the domestic heating market they have gone from being an innovative but occasional curiosity to being a quaint and occasional curiosity without ever passing through a phase of general popularity.

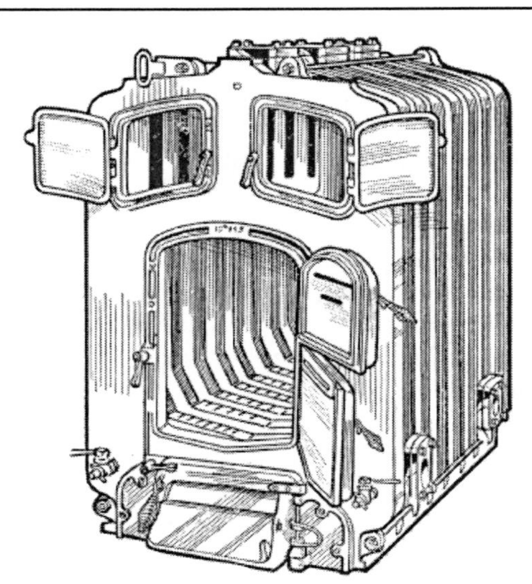

Figure 1.3: Illustration of a boiler made up of cast-iron sections bolted together with gaskets between them. This method of construction was used for large solid fuel fired boilers intended for use with gravity fed central heating systems. They were commonly installed in large public and industrial buildings during the first half of the 20th century but have never been popular in the domestic environment. From NC Pallot, *Principles and Practice of Heating and Ventilation,* George Newnes Ltd.London 1950.

This is largely down to the inconvenience and pollution associated with solid fuel use. The comparatively high power required of the boilers of central heating systems means that if solid fuel is used, automatic feed or regular stoking are needed. This increases the cost when compared to systems fired by oil and gas. Added to this the clean burning of oil and gas compared to solid fuel brings considerable environmental advantages. Much of the progress that has been made in cleaning up the air of our towns and cities has been possible because gas and oil burning heating systems, mainly central heating, have allowed a massive reduction in the amount of coal burnt. The spread of central heating was closely linked to this change in fuel use and there has never been a time when solid fuel fired central heating systems were widespread in homes.

The main fuels used for central heating have been gas and oil. Gas has dominated the market in the United Kingdom whereas oil has been the principal fuel in North America and much of Europe. There are various reasons for this distinction. Great Britain has had a ready supply of gas from an early date because it was produced from locally mined coal which was the principal energy source for two centuries. Furthermore the relatively small size and high population density of the British Isles made it economic to connect most houses to centrally piped gas.

Prior to 1967, coal gas was used for the mains gas supply in the United Kingdom. It was produced in numerous urban gasworks around the British Isles by heating coal in a reducing atmosphere. Its main constituents were hydrogen, methane, and carbon monoxide[3].

Early developments in the use of coal gas were dominated by its application to lighting rather than heating. When it burns after flowing out of a simple hole in a gas pipe it does so with a bright yellow flame, brighter than that of natural gas. This was the basis of its early use for gas lighting. Gas mantles did not become available until 1887.

A few experimenters worked on the possibilities of gas lighting during the 18th and 19th centuries. The Scotsman William Murdoch was lighting his cottage and office in Cornwall by gas in 1792 and extended the system to a factory in Birmingham by 1803. These were isolated occurrences but by 1823 the "gas light and Coke Co" was lighting 215 miles of London's streets by piped gas.

[3] Carbon monoxide is poisonous. During the coal gas era many people committed suicide by putting their head in a gas oven with the gas turned on but not lit. This no longer works as natural gas contains virtually no carbon monoxide. Carbon monoxide poisoning remained a popular means of suicide because it was contained in significant amounts in car exhaust. Even this no longer works because modern clean burning engines with catalytic converters produce negligible amounts of carbon monoxide.

Although gas lighting was inefficient, it consumed gas much more slowly than heating systems. It was feasible, though inconvenient, to store gas in containers at atmospheric pressure for lighting. The increased rate of burning required by heating systems meant that it was impractical to use this means of storage and viable heating systems had to await the arrival of piped mains gas.

Mains gas became available to a few domestic and municipal users during the first half of the 19th century. Most people still depended on solid fuel heating and oil lighting but there were enough houses with access to mains gas to prompt commercial development of gas appliances. To begin with these were primarily aimed at cooking. The date of the first use of gas cooking in a domestic environment is uncertain but an early credible report dates from 1830 when James Sharp demonstrated a cooking range that he manufactured and sold from his own home in Northampton. Gas cooking technology had evolved to the point that cookers were of a recognisably modern layout by 1852 but they remained a novelty until a much later date because of the small proportion of the population who had access to mains gas.

A critical step in the development of gas heating systems was the mixing of gas with air before burning. Prior to the 1840s, all gas lighting and heating systems worked by allowing gas to pass out of a hole in a pipe and burning it in the air. Used in this way coal gas burns with a bright yellow flame but not much heat. During the 1840s, systems began to emerge where gas and air were mixed before being burnt. This led to a blue flame which produced very little light but much more heat than the yellow flames of non pre-mixed gas burners. There is some controversy about where the credit for this development lies. Peter Desdga and Michael Faraday both developed operational systems but the name most closely associated with pre-combustion mixing is Baron Bunsen. In 1855 he noticed that the flickering yellow flame of gas coming from a simple burner turned blue and became much hotter if a hole was made in the side of the tube upstream of the gas outlet. He found that such a hole allowed air to flow into the tube rather than gas to flow out. This development made possible practical gas fires which used pre-mixed gas flames to heat ceramic materials which in turn radiate heat into the room.

Gas mantles first appeared in 1887. Early models were fragile, expensive, and difficult to replace but by the mid 1890s technical advances in mantel manufacture had resulted in an efficient and dependable lighting system. As a consequence domestic gas lighting became very popular. Gas light piping was installed in many newly built houses. The gas mains distribution system expanded to accommodate this increased demand. The gas lighting era was quite short, but as gas gave way to electric lighting, its use as a fuel for heating and cooking appliances took off.

In 1966 the decision was taken to change the entire United Kingdom mains gas supply to natural gas. This was in the light of the discovery of large reserves of natural gas under the North Sea. Natural gas is almost 100% methane and coal gas burning appliances had to be adapted to burn it cleanly and efficiently. Over the 10 years following the decision to convert the country, 34 million gas appliances were changed over!

From the point of view of the user, the main developments in the last thirty years have been in the hot water system rather than the space heating. Domestic systems installed before around 1980 use a pump to drive hot water round the space heating circuit but use convection to divert water from the space heating circuit through a copper coil heat exchanger located inside a hot water tank. The tank is kept full of water at a constant pressure by a float valve controlled cistern usually located in the attic. This gives a limited supply of hot water that is particularly inconvenient when running more than one bath consecutively.

The combination boiler avoids this problem by heating water directly and not storing hot water in a tank. Early models dating from the 1980s were unreliable and only allowed a low flow rate of hot water but modern systems perform much better in this regard and combination or "combi" central heating systems remain a competitive and popular choice. Increasingly popular, though, are the mains pressure hot water systems of un-vented cylinders and thermal stores. These are not a new innovation. They have been used for many years in public buildings but the increasing demand for high performance hot water systems in the domestic environment has prompted their development on a smaller scale to supply this market. At the time of writing, mass storage and un-vented systems are the best performing hot water systems available. The cost of installation and space required are higher than for combi systems but the differential cost is no longer great and they are an increasingly popular choice.

Global Warming

Throughout the 20th century the efficiency, convenience, and ease of use of central heating systems steadily improved. The regulations applicable to them also steadily evolved with the principal aims of ensuring safe operation and minimising annoyance to neighbours and users from noise or smoke production. Global warming began to rear its head in the 1980s as a political issue and at the 1997 Kyoto summit an agreement was reached across Europe and North America on controlling carbon dioxide emissions. In the wake of this a series of new regulations were introduced not aimed at annoyance or safety but at minimising the amount of fuel necessary to heat buildings. These regulations extended beyond the heating appliances themselves to the installation of insulation, windows, door frames etc. With regard to the performance of the central heating systems certain previously common practices were no longer permitted in newly installed systems and users and installers were encouraged to upgrade existing systems where feasible.

Boiler performance had to be improved on a rolling basis. The effect of this was as of April 2005 all gas boilers installed have to have an A or B SEDBUK[4] rating – which in effect means they have to be condensing[5]. Other high efficiency technologies that could potentially comply with efficiency bands A and B are on the horizon. Principle amongst these is micro combined heat and power (CHP). It refers to systems that burn fuel to produce both heat and electric power. In effect the heat recovered by the condensing system is recovered by CHP but converted into electricity which may or may not be used to heat water. CHP systems are available for industrial sized applications. Scaled-down versions for use in the domestic environment known as "micro CHP" are under development.

All systems now have to be fully pumped making the old gravity fed systems and partly pumped systems obsolete. Where possible such systems should be upgraded to fully pumped systems.

Hot water cylinders have to be insulated to higher standards than previously. Newly installed cylinders have to be factory insulated. The older tie on jackets are no longer permitted except to lag cylinders that are already installed.

Central heating control systems should use appropriate zoning. This is taken to mean that areas of a building with reasonably different heating requirements should be zoned. In practice this will mean separate zoning for the upstairs and downstairs of houses. The zoning will not be required in most flats or open plan design dwellings. These zones should each be independently thermostatically controlled and the system should incorporate a suitable timing device.

Provision should be made to prevent boiler interlock. Basically this means that the control system should switch the boiler off if neither the

[4] **SEDBUK** stands for Seasonal Efficiency of Domestic Boilers in the UK. It is due to be superseded by an expected European directive on boiler efficiency. It grades boilers A to G with efficiency ranges in % of >90, 86-90, 82-86, 78-82, 74-78, 70-74, and <70 respectively.

[5] Condensing boilers improve efficiency by condensing water vapour from the exhaust of the fire.

hot water nor the space heating is calling for heat rather than allow the boiler to be controlled by its own thermostat and simply divert heat to those areas calling for it.

Although these regulations were introduced to minimise carbon dioxide production for environmental reasons, they also had significant beneficial effects from the point of view of central heating system users. They have brought about an improvement in the comfort, convenience, and efficiency of systems. Many of the practices that have been disallowed were nothing more than poor engineering and short-term false economies, with which the uses of the system had to live for years!

It is ironical that while legislation stemming from the Kyoto agreement has improved the quality of our central heating systems it will probably achieve virtually nothing towards its aim of controlling global warming. Increasing evidence is suggesting that while global warming may be in part down to human activity, it is no longer feasible for mankind to significantly influence the timing or extent of its ultimate course.

Chapter 2
Heating Circuits

The "circuit" of a central heating system refers to the route taken by water as it leaves the boiler, passes round the building, and returns to the boiler or flows out of taps. There are usually two components to this circuit. One of them carries water around the radiators and the other heats water for the taps. By convention the circuit to the radiators is known as the "space heating" and that to the taps the "hot water". The hot water system involves two circuits. The primary hot water circuit uses the same hot water as flows in the space heating circuit and passes it through a heat exchanger in the hot water tank. This indirectly heats the water in the secondary hot water circuit so that there is no mixing of water that has been through the boiler with that going to the taps.

Systems can be un-pumped, partly pumped or fully pumped. Un-pumped systems rely on thermal convection currents to drive hot water around both the space heating and primary hot water circuits. Partly pumped systems use thermal convection for the primary hot water circuit but pump water round the space heating circuit. Fully pumped systems pump water around both.

Systems can also be either single or two pipe where the distinction refers to the way the space heating circuit operates. Single pipe systems have only one pipe carrying water round all radiators (figure 2.1). Two pipe systems have flow and return pipes to the radiators (figure 2.2). At some time or other every combination of pumped, partly pumped or un-pumped and single or two pipe systems have been used but the commonest to be encountered now are single pipe un-pumped old installation in large buildings, two pipe partly pumped systems which were the standard installation in domestic houses for many years and two pipe fully pumped that are the standard now.

If you are installing new central heating in a dwelling, the best type will almost always be a two pipe fully pumped system. This means that the flow of water to both the space heating and hot water circuits will be driven by pumps and that the radiators will have separate pipes conveying the water to them (called flow) and back from them to the boiler (called

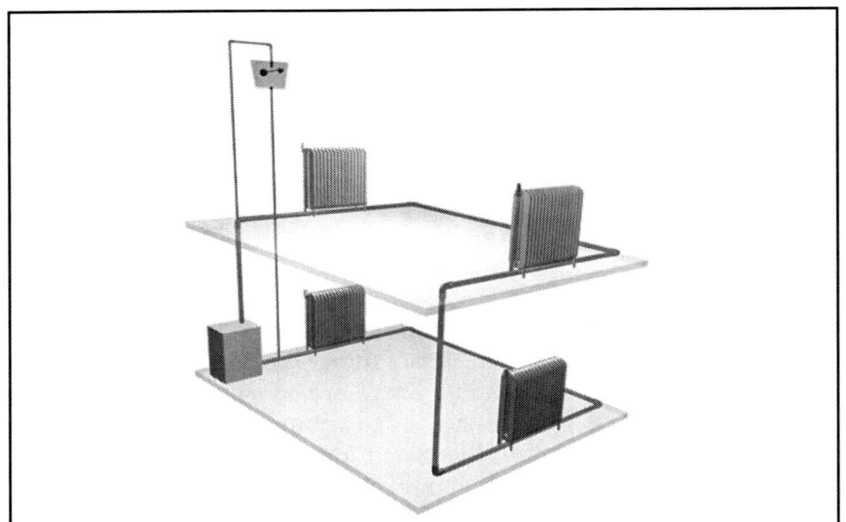

Figure 2.1: Single pipe central heating system showing the vent, cistern, and cold feed. The mains cold water supply to the float valve in the cistern is omitted for clarity.

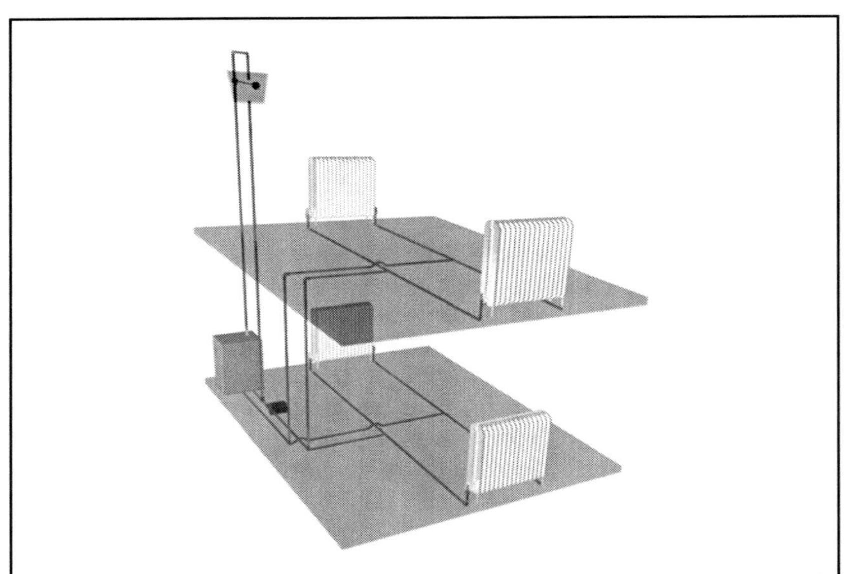

Figure 2.2: Two pipe pumped central heating system. The vent, cistern, and cold feed are shown. The mains cold water supply to the float valve in the cistern is omitted for clarity.

return). Two pipe fully pumped systems are the most efficient, most rapidly responsive and nowadays the most economic. Domestic central heating systems have evolved through several types and other systems are still installed in larger buildings so they are frequently encountered, necessitating a knowledge of their characteristics and construction.

Water pressure around a central heating circuit

Hydrostatics

Hydrostatics is the study of the behaviour of liquids that are not moving, or in the present context, of water in a central heating system that is stationary rather than being driven around by pumps or convection.

Pressure is defined as force per unit area. Several different units are in use to measure pressure. The standard scientific unit is the Pascal (Pa) which is equal to 1 N[1] per metre squared. This is a relatively low pressure. Atmospheric pressure is equal to 101,325Pa or approximately 1 bar (100kPa) where 1 bar = 100,000Pa and 1kPa = 1000Pa.

The basic equation of hydrostatics is that the pressure at a point in a fluid is equal to the height of fluid above that point times the density of the fluid times the acceleration due to gravity[2]:

$$P=h\rho g$$

Where P is pressure in Pa, h is height in m, ρ is density in kg/m^3 and g is the acceleration due to gravity ($9.81ms^{-2}$). Hydrostatic pressure depends on these factors only and not on width or layout of pipes.

In the case of water $\rho = 1000kg/m^3$. As the acceleration due to gravity is almost constant everywhere on the earths surface this leads to a convenient way of reckoning pressures as the height of a column of water. From the formula above 10m of water = 98,100Pa which in practice is close enough to atmospheric pressure (101,325) to be regarded as the same. The height of a column of water over a particular pressure varies slightly with the temperature as the density of the water changes.

Hydrostatic pressures are usually expressed relative to atmospheric pressure rather than to a pressure of 0. This means that pressures below atmospheric are referred to as negative and those above as positive. This is convenient as the difference between pressures rather than absolute pressures are important in determining flow.

[1] The SI unit of force is the Newton (N). This is defined as the amount of force required to accelerate an object with a mass of 1 kg by 1 m per second every second. One N is a relatively small force. It is enough to support a weight of 102 g.

[2] The acceleration due to gravity is equal to $9.80665\ m/s^2$. It is rounded above to 9.81 but 10 is accurate enough for most purposes.

The hydrostatic equation means that pressure varies in a central heating system depending on the height of components. It is convenient and common practice to reckon pressures in central heating circuits in m of water. In most installations the boiler and pump are installed on the ground floor. In a two-storey house the water pressure in the boiler will be around 5 m higher than in the attic.

Hydrodynamics

Hydrodynamics is the study of the behaviour of liquids in motion, and in our present context of water flowing around a central heating circuit. The basic equation for Hydrodynamics gives the rate of water flow through system components such as radiators, valves, boilers and pipes. The rate of hot water flow through a component can be calculated from its resistance to flow and the difference in pressure between its ends. A simple approximation that is good enough is that the flow rate is directly proportional to pressure across a given resistance. This is analogous to Ohm's law in electric circuits where current flow through a resistor is equal to the voltage difference between its ends divided by its resistance.

The reckoning of pressures in a central heating system becomes considerably more complicated once a pump is installed and flow has to be taken account of. Flow related pressures are referred to a pressure "datum". This is a particular point in the central heating system where the pressure is fairly constant. In an open vented system it is the point where the cold feed is connected to the central heating circuit. It comes down from a cistern in the attic in which the level of water is kept constant by a float valve. The water is cold with little temperature variation and so the pressure is fairly constant. The situation in a sealed system is different as the pressure is not constant but the concept of the pressure datum is of less value.

The calculation of exact pressures and flow rates is complicated because the output of a pump varies considerably with the resistance of the circuit it is pumping water through as illustrated in figure 10.2 meaning that pressures are dependant on flow rate which is in turn dependant on pressures! It is not necessary to calculate exact pressures or flows for installation purposes. The most important considerations are the maximum and minimum pressures that may arise and they are much more easily reckoned.

An important property of complete circuits is that if the pressure differences between the ends of all the components in a continuous circuit including pumps, radiators, valves, pipes, etc. are added together then the result will always be zero. We're now in a position to examine what impact pumping will have on pressures within a central heating circuit and consequently to determine what adverse events may be caused by pump related pressures. This will enable us to determine the optimal position in the circuit for the pump.

When a pump is installed in a circuit the overall datum pressure of the circuit remains as for static pressure but with the pump running a pressure difference will appear from one end to the other depending on the flow rate and setting. This works like a floating isolated electric circuit containing a battery. Figure 2.3 illustrates the effect of turning on the pump. The most extreme influence the pump can have on pressure is to add or subtract the maximum pressure the pump can raise to or from the static pressure. In a sealed system the static pressure is well over the maximum pressure raise of the pump so the pump can never make a local pressure negative unless the system pressure falls, but this is possible in an open-vented system.

Figure 2.3 shows a simple central heating circuit that includes the pump, radiator, boiler, cold feed and vent. The columns at each point show the pressures relative to the datum. As an example assume a flow rate of 0.3 l/s with the pump on a medium setting. Figure 10.2 shows that the pump pressure rise will be about 4.3 m. If the boiler is a low resistance old-fashioned cast-iron one which is in good condition and the radiator valves are almost closed the pressure raised by the pump will almost all appears across the radiator with very little across the boiler.

The principal pump related problems that we want to avoid are air in-drawing and pump-over. Both of these relate to the pressure at the bottom of the vent pipe. Pump-over occurs when the pressure at the bottom of the vent pipe rises and water is driven out of the vent into the cistern. This will result in circulation of water down the cold feed, up the vent pipe and back into the cistern. This is undesirable because the pouring of water out of the vent pipe through the air into the cistern leads to increased oxygen dissolved in the water and accelerated corrosion. Air in-drawing occurs when the pressure at the bottom of the vent is lowered so much that all the water is sucked out of the vent and air enters the system. This leads to rapid failure because of air locks in pipes and because pumps are ineffective at pumping air. The way to avoid these problems is to ensure that the cold feed and vent are placed close to each other in the circuit and are not placed on either side of the pump, boiler, or radiators.

Un-pumped systems

When the temperature of water changes, it expands or contracts and consequently its density changes. It has a maximum density of 1kg/l at 3.98°C. Cooling or warming water from this temperature makes it expand and become less dense. The relationship between temperature and density is quasi-parabolic rather than linear as shown in figure 2.4. This expansion is the origin of the convective currents that drive hot water round un-pumped systems. Less hot water will fit in the same volume as cold water so hot water is less dense than cooler water. If a circuit with parts at different heights such as that in figure 2.1 contains hotter water on one side than the other, this density difference leads to a pressure difference. This pressure difference in Pascals per m height is equal to the difference in densities at the two temperatures (figure 2.4) times the acceleration due to gravity of 9.81m/s². A typically boiler in such a system might raise the temperature of water passing through it from 56 to 70 °C leading to a pressure difference of 72.1 Pa/m. In a large building the radiators on the second floor might be 10m above the boiler giving an overall pressure of 721 Pa. This is the hydrostatic pressure of a 7.21 cm high column of water and is very little by comparison with the pressure pumps can raise. Consequently un-pumped systems require low resistance wide diameter pipes to convey the water around. Typically the circuit pipes are 50 to 100mm across. For many years the above calculations were the nuts and bolts of central heating design[3], thankfully no longer.

Un-pumped single pipe

Single pipe systems (figure 2.1) use one large pipe that carries the hot water round all the radiators. The system is quite easy to recognise as usually the single pipe lies above floor level underneath the radiators which are of a cast-iron construction with small feet that stand on either side of the large pipe. Flow through the radiators is driven by convective currents. Figure 2.5 shows how radiators in single pipe systems are connected up. The arrow shows the direction of flow through the single pipe. This direction is from the hot flow from the boiler, round the circuit to the cooler return. It is usual for the inlet flow to radiators to be connected to the single pipe upstream of the outlet from the radiators as shown in figure 2.5. This is not critical and the system will still work if the connections are reversed. One of the connections is to the top of the radiator and the other to the bottom. This is because as water within the radiator cools it sinks. It will flow out of the lower connection and draw water into the higher one. The radiator will still work if both connections are at the bottom but flow will be more sluggish. From the figures mentioned above it can be seen that the pressure driving water round a

[3] See J. R. Kell's *Heating and Air-conditioning of Buildings*, Architectural Press, London 1979 ISBN 0851392881.

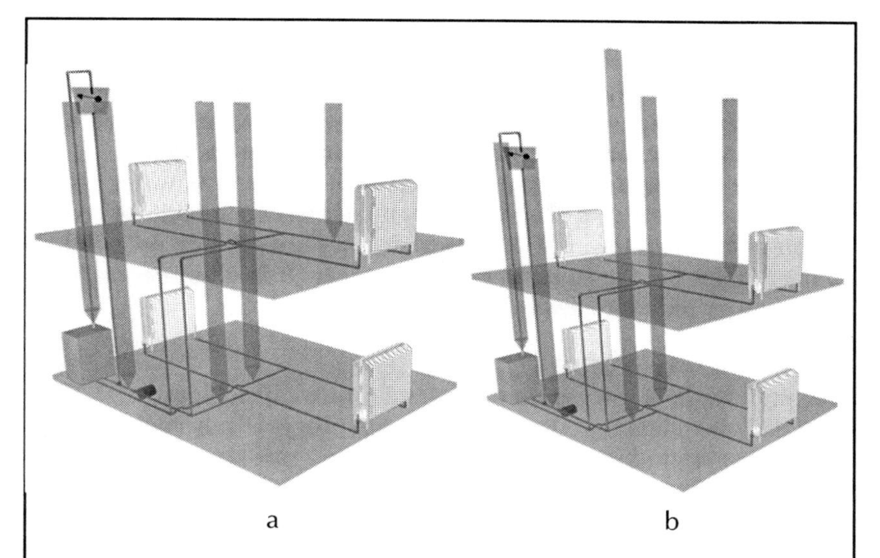

Figure 2.3: Illustration of the effects of pumping water on pressures in an open vented system. The perspective of the pictures is arranged so that the water level in the cistern is in a horizontal plane at eye level. Figure 2.3a illustrates the static situation when the pump is switched off. The hydrostatic head of water is the same at all points. The pressure at the various points in the system depends only on their height above the datum - the point at which the cold feed joins the rest of the circuit. Figure 2.3b shows the situation when the pump is running. The effect is to produce a pressure difference from one side of the pump to the other with the inlet of the pump at a lower pressure that the outflow. The pressure then falls across all other components of the system so that the pressure rise across the pump is exactly equal to the total pressure fall around the rest of the circuit. The pressure is highest immediately after the pump. It falls somewhat after the radiators. The pressure in the upper floor return piping is considerably lower than on the lower floor but this is because of the greater height of the upper floor. The resistance of the flow piping is small and hydrodynamic[4] pressure differences between various points on the flow side are similarly small with the lowest pressure being on the return side of the boiler. The pressure then falls further across the boiler. It is at this point that the cold feed is connected in this diagram. The cold feed pressure is constant and acts as a datum against which all others are measured. In this situation, because the cold feed is immediately behind the pump, it has the lowest pressure in the system and all others are above it though this is not the only possible arrangement.

[4] Hydrodynamics is the effect of moving water. The differences between figures 2.3a and b are hydrodynamic pressures.

radiator connected to a single pipe system is extremely low at around 3 to 10 Pa. The pipes running to and from the radiator and the channels within it must be quite large. The radiators made in the first half of 20th-century were designed to be used in this way. They were large cast-iron structures with generous channels within them (figure 2.5).

Single pipe systems have the drawback that as the water moves round the circuit it gets progressively cooler after each radiator so that radiators installed towards the return end of the circuit do not produce as much heat for the same size as the others. It is often necessary to install larger radiators towards the cooler end.

Un-pumped two pipe systems

Two pipe systems use separate flow and return pipes (figure 2.2). They have the advantage that the water does not progressively cool as it goes round the system but remains at roughly the same temperature in the flow pipe so that radiators do not need to get bigger towards the return end of the circuit. The disadvantage is that installation costs are higher because two pipes have to be installed instead of one, both of which must be quite large. The pressure driving water through the radiators is considerably larger than with the single pipe system because the radiators form part of the entire convective circuit rather than a number of individual sub-circuits.

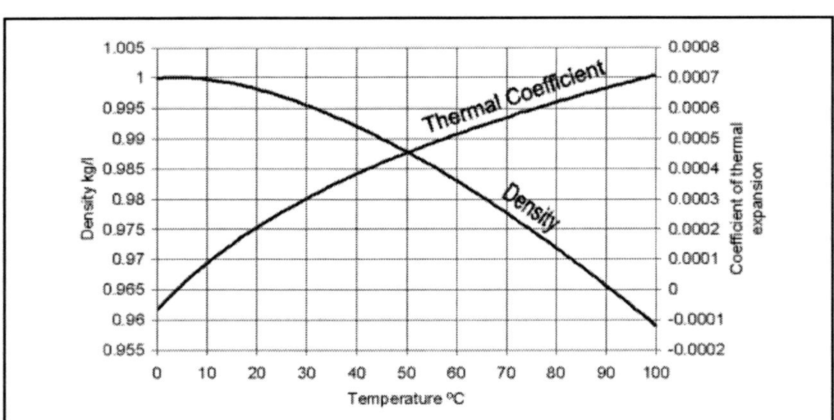

Figure 2.4: Graph showing how the density and coefficient of thermal expansion of water change with temperature. The graph can be used to determine what the change in volume of water for a specific temperature change will be. Consider a volume of water v that changes from temperature t1 to t2. Read off the graph the density for t1 (ρ1) and t2 (ρ2). The volume at t2 will then be vρ1/ρ2. At 3.98 °C the density = 1 and the thermal coefficient = 0.

Un-pumped systems have many disadvantages which make then unsuitable for domestic settings. The large pipe diameter necessary means that the radiator circuit contains considerably more water than it does on more modern systems. This gives a substantial heat capacity so it takes a long time to heat up and cool down. Coupled with the slow flow rates of convective currents this leads to poorly responsive systems. The slow turn on time is uncomfortable and the lingering turnoff time leads to considerable heat loss when the building is not in use.

A further disadvantage of un-pumped systems is that they perform poorly when installed in a single storey building and cannot heat rooms below the level of the boiler unless they are included in a single pipe circuit that extends considerably higher above, than it falls below, the level of the boiler. This is because convective currents require that hot water rises towards the radiators and falls away from them. There is little potential for this rising on a single storey unless the boiler can be accommodated on a floor below such as in the cellar.

Most convective systems are now quite old. The large diameter of the pipes involved generally meant that they had to be built out of steel for economic reasons. This leads to a problem of internal pipe corrosion but inhibitors (chemicals added to the water) largely correct this. On the whole these old systems were very well built and have proved to be particularly durable.

Figure 2.5: Radiator connection in a single pipe system. The inflow to the radiator is normally connected at the top and the outflow at the bottom. Ideally the inflow should be connected to the upstream side off the large pipe as shown in this figure but the system will still work if it is connected to the downstream side.

The low circulating pressures of un-pumped systems means that they will only work with low resistance boilers. These tend to be old cast-iron designs which are not as efficient as their modern high resistance counterparts.

Un-pumped systems have their advantages. They are very simple and many can operate without an electricity supply. This makes them reliable and proof against electrical failures. Many are still in operation after more than 50 years. It is doubtful whether the same will be true for contemporary systems though this is more a comment on standards of manufacturing and workmanship than on the systems themselves!

A further advantage of un-pumped systems is that they are almost completely silent. The principal causes of noise from central heating systems are pump vibration and the effects of the high pressures that pumps develop. Without these the only noise produced is that of the boiler itself.

Alterations to un-pumped systems

It may be necessary to add or remove radiators to un-pumped systems or to route the circuit round new parts of an old building. Techniques of working with the steel pipe these systems tend to be formed from are described in chapter 7.

It is fairly straightforward to add a radiator if a pipe is passing through the room. One side of the radiator is best connected at the top (ideally on the flow side of the pipe) and the other at the bottom and the system will work even if no controls are fitted. Control valves can be fitted but it is necessary carefully to select ones that will work with the very small pressure difference. These are usually specialist products designed particularly for convective central heating systems.

Ideally, if a radiator is added to a single pipe system, another radiator should be removed somewhere else otherwise the cooling tendency will be increased and the radiators further round the circuit will become cooler. In practice this effect is slight for a single additional radiator.

Adding a radiator where there is not a passing pipe is more difficult. It is possible to add a radiator on the floor above the pipe and run pipes to and from the radiator from the pipe on the floor below. In general such a radiator would run slightly hotter than the others close to it in the circuit because the greater height above the pipe means larger convective currents. This leads to difficulty in balancing the radiators but in practice un-pumped systems tend to be either on or off and issues such a balancing radiators or adding control valves are often of little relevance unless being used to correct an over heated room.

Rerouting a single pipe round new parts of the building is also possible. As a general rule any major restructuring of un-pumped systems is not likely to be economically competitive compared with a replacing the whole system with a modern pumped one.

The most obvious means of rerouting un-pumped systems is to splice into the single pipe and run it around the additional rooms. Additional pipe runs of this type should be kept as short as possible to minimise the increase in resistance to water flow offered by the whole circuit and they should be confined to one floor unless replacing existing transitions between floors. Running the single pipe up and down again is likely to adversely affect the performance of the whole system.

An alternative approach to running the single pipe up and down between floors is to add in a two pipe extension. This works as would a large radiator except that the two pipes can feed several radiators in the extension. If this kind of extension is used the pipes connecting the radiators to the circuit should be larger than for single radiators unless the two pipes come off a single pipe on the floor below the radiators, in which case convection in the extension will be enhanced.

Pipes sizing for gravity central heating systems is a complex issue that requires considerable design work. The best guide when altering a system is to use the same diameter of pipe for additional radiators as is used for existing ones and if the single pipe is being extended to use a pipe of at least as large a diameter. If a two pipe extension is being added it should be fed by pipes of intermediate diameter between the single pipe and existing radiator pipes. Twenty two or 28 mm would be adequate in most cases[5].

Partly pumped systems

Partly pumped systems use an electric pump to drive water round the radiator circuit but not round the primary hot water circuit. For a long time this was the standard domestic installation. It is possible to have single pipe or twin pipe partly pumped systems but in practice single pipe pumped systems have rarely been used in domestic environments. They are encountered in larger public and industrial buildings. Electric pumps raise far higher pressures than the convective pressures discussed above and it is possible drastically to reduce the diameter of pipes delivering water to the radiators. This significantly reduces the cost and complexity of installing the system but also reduces the total

[5] Pipe sizing calculations for gravity systems involves not only assessing flow rates but also heat radiated from the surface of pipes. Details can be found in J. R. Kell's Heating and Air-conditioning of Buildings, Architectural Press, London 1979 ISBN 0851392881

volume of water contained within the circuit. Consequently the total heat capacity of the system, and the time it takes it to heat up and cool down are reduced. The much greater speed in metres per second (as opposed to the flow rate in litres per second) of pumped water adds to the improved response rate of these systems.

Convective hot water

Partly pumped systems use no pump in the hot water circuit. In practice they are almost always "open vented" which leads to a number of complexities that are discussed under open vented systems below. In most cases the flow and return to the heat exchanger in the hot water tank are connected directly to the flow and return to the radiators coming from the boiler. This means the hot water is on whenever the central heating is on but it is possible to have the hot water on with the central heating off. The hot water tank has to be installed above the boiler, usually on the floor above, to give enough height for the convective currents to act. The supply of hot water is not dependent on electrical pumps which are prone to failure. In most domestic environments it is easier to cope with no central heating than with no hot water. The simple contingency of an immersion heater allows a backup hot water system but it will not work in the presence of a power cut!

Fully pumped systems

In a fully pumped system the flow of water from the boiler is pumped both around the radiator circuit and the primary hot water circuit. Most modern systems use one pump to do this. Electrically operated valves are used to divert hot water to the various parts of the system as demand changes. As a consequence much of the discussion of fully pumped systems revolves around the design and functioning of valves rather than pumps!

Three port valves

The most commonly encountered motorised valves in domestic central heating systems are three port valves. They come in two varieties: mid position valves and diverter valves. The difference is that diverter valves have a common connection and can switch it to either of two other connections but not both together. The mid position valves also has a common connection and can divert it to one or other of two connections but has an additional option to divert it to both simultaneously. Three port mid position valves are used in what's known as a Y plan circuit. Three port diverter valves are used in what is called a W plan circuit.

The Y plan circuit is now the standard installation. The W plan circuit was popular in the past and may still be preferred in some cases.

The W plan circuit

In this circuit when the control system request space heating only, the valve is moved to divert the boiler's output to the radiators. When the control system request heat to the hot water only, the valve's output is moved to the primary hot water circuit. The W plan circuit is hot water priority so that when heat is requested by both space heating and hot water the valve is moved to the hot water circuit only until its call for heat is satisfied when it is moved to space heating.

The Y plan circuit

The Y plan circuit works as does the W plan with the exception that when the control system is requesting both space heat and hot water, the valve mid position is selected and hot water flows to both the primary hot water and space heating circuits. The advantage of the system is that space heating is available while the hot water is being heated. This is important particularly when occupants arrive in a cold house where the heating and hot water have been off. The disadvantage is that this availability of space heating is at the expense of power to the hot water which takes longer to heat up.

It is a fairly easy matter to change a W plan to a Y plan circuit simply by changing the diverter valve for a mid position valve and changing the control circuitry. It may be possible to change the valve motor and not the whole valve. Before doing so it is sensible to ensure that more responsive central heating at the cost of less rapid hot water availability is what is actually wanted! Another option is to fit a control system which allows switching of the three port diverter valve between positions more frequently giving the same effect as a mid position valve.

Two port valve systems.

Two port valves are motor driven cocks that are either on or off. Systems based on to port valves are known as S plan. They work by zoning the radiators and hot water system. The common arrangement is to have three "zones": one for the hot water, one for the upstairs radiators and one for the downstairs radiators. The system requires one thermostat per zone. Each thermostat controls the two port valve for its own zone.

Choice of pipe size

The standard installation uses 22 and 15mm pipe. Common practice is to install 22 mm pipes as the core distribution system usually carrying hot water to each floor and run water round the different radiators with

15 mm pipes unless the house is particularly large in which case 22mm pipe should be run far enough along each floor that no 15mm run goes to more than 3 radiators. It is unlikely to be necessary to resort to 28 mm pipe unless the houses so large that an industrial boiler is required.

It is recommended that no more than three radiators be connected to a single run of 15 mm pipe and the capacity of water pumps limits the total number radiators that can be driven from most systems to about 12. Both these rules can be bent if necessary when adding additional radiators to an existing system but if a differential pressure release valve[6] is to be used the maximum number radiators and pipe runs recommended by the boiler and pump manufacturer should be more closely observed.

An increasingly popular system is mini or micro bore. Modern pumps can raise considerable water pressures allowing adequate flow through smaller pipes. Microbore pipes are usually between 6 mm and 12 mm in diameter with 8 mm and 10 mm being popular sizes. The small diameter means that it is usual to run only one radiator from one run of pipe. Systems are laid out with large feed pipes usually 22 mm in diameter fitted with manifolds into which numerous microbore pipes are connected. Each microbore run should be no more than 5 m in length. This leads to complications if an additional radiator is to be fitted because it means that there has to be a free manifold port to plumb it into. If there is not, an additional manifold has to be added which can be awkward and they are expensive!

Microbore systems are easy to install because the narrow diameter of the pipes makes them flexible and easy to manipulate round corners without complicated bending equipment. Their narrowness means water circulates rapidly[7]. This and the smaller volume of water in the system (lower heat capacity) makes radiators heat up quickly. A further advantage is that water in the pipes usually moves fast enough to prevent air locks from forming.

Whether to use 22mm and 15mm pipe or 22mm and microbore for a central heating system is a matter of personal choice to some extent. My own preferences is for conventional 22 and 15 mm pipe because

[6] These are valves connected between the flow and return sides of the radiator circuit. They limit the pressure difference between the two sides and prevent radiator noise.

[7] Rapidly here refers to the speed of water movement in metres per second rather than the flow rate in litres per minute. As an example 15 mm diameter copper pipe has an internal cross-section area of about 162 mm^2. If water is flowing through at the rate of 10 l per minute its speed will be 1 m/second. Compare this with 10 mm microbe or tubing with a cross-sectional area of about 70 mm^2 and the same flow rate the water will have a speed of 2.3 metres per second.

it is readily and cheaply available, is not complicated by the problems of running single pipes to radiators, is more resistant to knocks and bashes, and does not involve fiddling around with manifolds which are expensive and awkward. What can happen is you put a couple of pipes into the manifold and solder them. Then you do a couple more and heat the manifold again. This melts the solder in the first joints which now come apart!

Open vented and sealed systems

The above discussion refers to different ways in which water can be moved around a space heating and hot water system. A different classification refers to how a system is filled with water and how its static pressure[8] is controlled. The two options in this second classification are open vented and sealed systems. All central heating systems until quite recently were open vented but the more modern sealed systems have a number of advantages and have become increasingly popular. They are largely restricted to fully pumped systems. There is no reason why sealed systems cannot be partly pumped but they were developed during the period when fully pumped systems were the norm.

When water is heated it expands at a variable rate per °C given in figure 2.4. The average temperature of water in a central heating system typically ranges from around 10°C when the system is cold to around 75°C when hot. Reference to figure 2.4 shows that this leads to a 0.0255 or 2.55% expansion in the volume of water contained. Most domestic central heating systems contain between 20 and 50 litres of water which gives an expansion of roughly 0.5 to 1.3 litres. There is also expansion of the pipes and radiators themselves which slightly accommodates this but there remains an excess of water expansion. In open vented systems expansion is accommodated by an open cistern at the top of the building. In a sealed system it is accommodated by a special expansion vessel which allows a water chamber to expand by compressing a gas chamber.

Open Vented Systems

An open vented system uses a cistern in the attic to keep the circuit full of water at a constant pressure. As well as the cistern there is a vent pipe from which the system gets its name, the top of which is open to the air. Its purpose is to allow any air or gas that gets into the circuit to escape. The radiators all have air bleed screws as well for this purpose and the vent pipe's main function is to let air out of the top of the boiler

[8] Static pressures refer to the pressures in the system when the water is stationary.

to prevent a "dry fire"[9]. The open end of the vent pipe is also in the attic and is positioned above the cistern so that any water flowing out of it lands in the tank. Figures 2.1 and 2.2 show the basic layout.

Open vented systems take care to design and install. Thought must be given to the position of the open vent pipe relative to the cold water feed and pump so that air is not drawn into the system via the vent pipe and nor does water flow out of the vent into the cistern - a condition known as pump-over.

Pressures differences around the circuit

When a pump is used to drive water around the circuit it works by raising the water pressure beyond the pump. For all other components as the water flows through the resistances of the parts of the circuit, the pressure falls. This means that the pressure is greater on the inflow side of a particular component than it is on the outflow side. This is illustrated in figure 2.3.

During normal operation the float valve in the cistern keeps the height of water in the cold feed circuit constant. The point where the cold feed joins the circuit is used as the pressure datum[10] from which other pressures around the system are reckoned. Pressures above the datum are referred to as positive and those below it as negative.

We now come to the complex question of the order in which to install the vent pipe, cold feed, pump, primary hot water circuit and space heating circuit in open vented systems. There are several constraints which limit the order in which these are connected. When these are respected they leave 3 commonly used arrangements. We will consider the constraints and the reasons for them first.

Simple boilers consist of a fire box surrounded by a water jacket. Such boilers are rarely installed now but for many years they were the mainstay of domestic central heating. To ensure that the water jacket is entirely full of water it has to be vented at the top. As convective currents make water rise as it is heated, hot water is drawn off from the top of the water jacket so the vent pipe must be at or close to the point of out flow from the boiler, in other words on the flow side. A specific connection on the boiler is frequently provided for the vent. In these cases no other

[9] A dry fire refers to the situation when the boiler fire is lit but the water jacket contains air rather than water. This will destroy the boiler in minutes or less because of the rapid and gross overheating of the jacket or heat exchanger that occurs. Considerable pains are taken in all central heating systems to avoid a dry fire.

[10] A "datum" is a convenient fixed reference point. In this case the bottom of the cold feed is chosen as a datum because this point has a constant unchanging pressure, a property not shared by any other point in the circuit.

connections can be made to the flow pipe between the boiler and vent. If the cold feed is connected upstream of the vent it must therefore be on the other side of the boiler.

A further constraint is that the difference in pressure between the vent and cold feed connections cannot be that great. If the vent pressure is lower than the cold feed by a significant amount there is a risk that air will be drawn into the system through the vent. On the other hand if the vent pressure is higher than the cold feed there is a risk of pump-over. The pressure difference needed to provoke air in-drawing has to be 5 metres or so in order to suck air all the way down the vent pipe from the attic. The pressure difference needed to provoke pump-over is much more modest as the open end of the pipe is rarely more than a metre above the tank. Because of these constraints the pump should not be installed between the cold feed and vent.

Another constraint is that water flowing in the primary hot water circuit should not pass any connections to the space heating circuit on its way to or from the hot water cylinder. This is to avoid reverse circulation faults which are caused by convective flow around the space heating circuit when it is supposed to be switched off. Reverse circulation makes radiators heat up unbidden when heat is not required and space heating is meant to be off.

The pump should be placed so that both the space heating and primary hot water circuit are on the same side of it. When the pump is in operation some of its output flows round the primary hot water circuit and some round the space heating circuit but all of the pumped water has passed through the boiler so is hot. If the pump is placed between the primary hot water circuit and the space heating circuit then all of its output will flow through the space heating circuit but its input will be divided between water that has passed through the water jacket of the boiler and water that has gone backwards through the primary hot water circuit. This latter component of the pumped water will on the one hand be cooler than water coming from the boiler and on the other hand will cool water in the hot water tank rather than heating it. The pump should be positioned on either the flow or the return side but should not be positioned between the flow or return connections to the primary hot water circuit and space heating.

These constraints leave a limited number of possible layouts that are shown in figures 2.6 to 2.9.

The 3T layout – Figure 2.6

Each connection to the main flow and return pipes is known as a "T" because they are made with soldered or compression T unions. The 3T

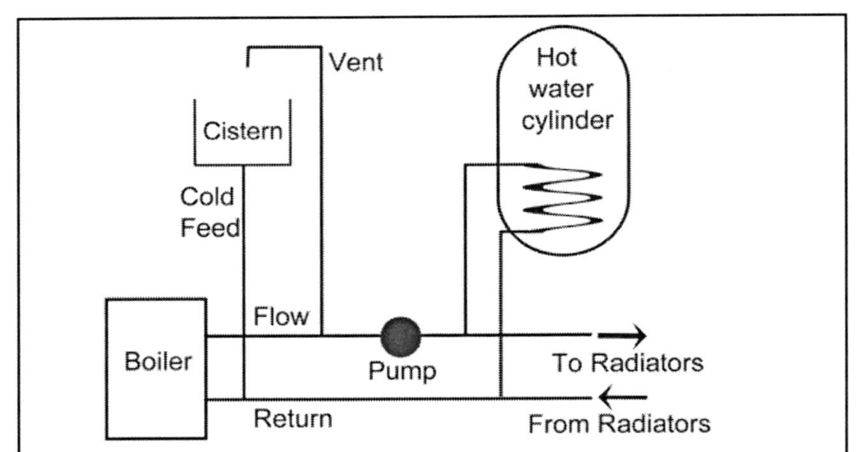

Figure 2.6: The 3T layout of open vented central heating systems. This circuit arrangement is suitable for low resistance boilers. These tend to be floor standing boilers with cast-iron water jackets. More modern boilers, particularly gas-fired boilers, are high resistance and not suitable for use with this circuit.

rule refers to the order in which the three connections on the flow and return sides are made. On the flow side they are in order, when moving away from the boiler; vent, primary hot water flow, and space heating flow. On the return side moving toward the boiler they are space heating return, primary hot water return and cold feed.

This arrangement shows the conventional place to connect the cold feed. It is upstream of the vent so the pressure in the cold feed will always be higher than the vent and there is no risk of pump-over. The amount by which the vent pressure is lower than the cold feed pressure depends on the pump flow rate and the resistance to water flow through the boiler's water jacket and the piping between the cold feed and vent. Old-fashioned cast-iron floor standing boilers offer very low resistance to water flow through the jacket so the negative pressure in the vent is modest and the drawdown of water in the vent pipe is of the order of a metre or so at most. This gives a larger margin of safety to avoid air in drawing. Air in drawing can occur if the boiler jacket or associated piping become heavily obstructed by lime scale or corrosion but this is unlikely to occur in a well maintained and normally functioning system. This arrangement is a good choice for low resistance boiler and was the standard installation in the past.

Figure 2 .6 shows the pump on the flow side between the vent and the primary hot water flow. The pump can also be placed on the return

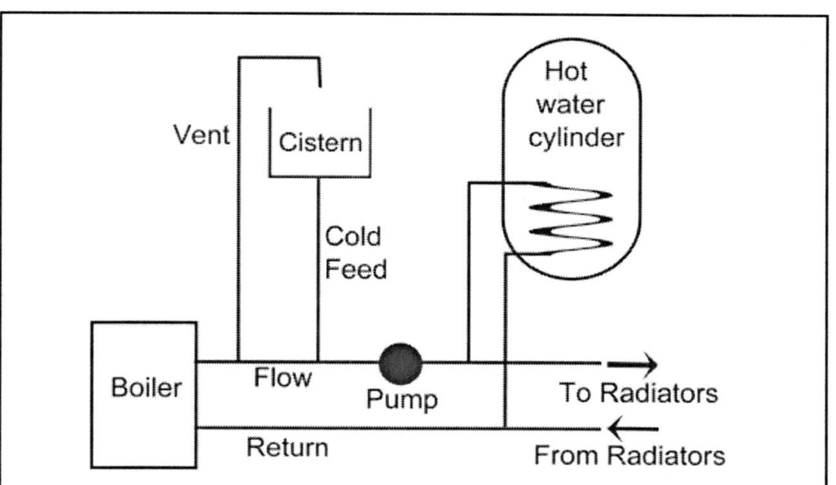

Figure 2.7: The 4T 2T layout of open vented central heating systems. This circuit is suitable for modern high resistance boilers. The order of connection of the cold feed and vent is reversed from the 3T layout. This means that the risk of air in-drawing is replaced with a risk of pump-over. To minimise this risk the pressure difference between the two connections should be the smallest possible. They should be placed not more than 150 mm apart on a 22 mm or greater diameter flow pipe.

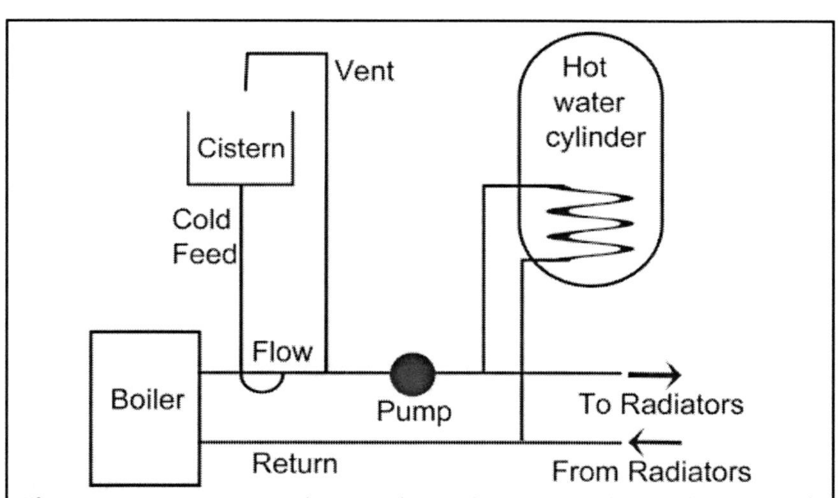

Figure 2.8: A variant on the 4T 2T layout that is more tolerant of pressure differences between the cold feed and vent. In order to prevent air bubbling up the cold feed pipe it is connected to the bottom of the flow pipe. This system is better if a longer distance between the connections of the two pipes is necessary.

side between the primary hot water return and the cold feed but putting it on the flow side means that radiators lie between the pump outlet and cold feed so will be at a positive pressure. This has the advantage that any leaks are visible because water leaks out rather the air leaking in.

The 4T 2T layout - Figures 2.7 and 2.8
Modern boilers, particularly those that burn gas, do not use a conventional cast-iron water jacket but a matrix heat exchanger of fine channels that contain a small volume of fast moving water which is heated rapidly. These boilers have a high resistance to water flow and are not suitable for use it in the 3T circuit because the high resistance causes a high pressure drop across the boiler and a significant risk of air in drawing down the vent. The high speed of water movement in the matrix prevents air from collecting within it so the vent does not need to be placed immediately at the outlet from the boiler. This allows the 4T 2T circuit to be used. The vent is connected upstream of the cold feed because otherwise any air leaving the boiler would pass the cold feed first and bubble up towards the cistern. This is fairly harmless but could cause an air lock if the pipe to the tank goes round an inverted U. As the cold feed is now downstream of the vent there is a possibility of pump-over rather than of air in-drawing. The pressure difference necessary to provoke pump-over is much smaller than that necessary provoke air in

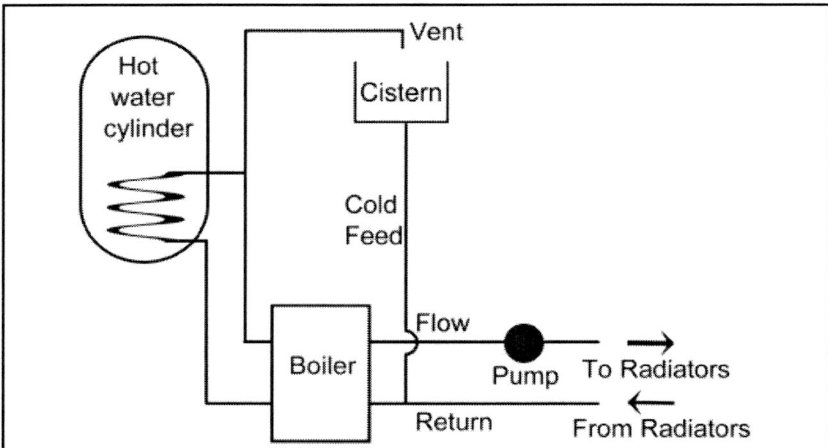

Figure 2.9: Some boilers have separate connections for the hot water and space heating. When installed in partly pumped open vented systems this circuit is used. The problem of reverse flow is minimised because all water entering the pump has passed through the water jacket off the boiler, even if it has gone backwards through the primary hot water circuit.

drawing. For this reason special care must be taken to ensure that the pressure difference between the two connections is as small as possible. They should both be connected to a flow pipe at least 22mm in diameter and the distance between the two connections should not be more than 150mm.

Another possible approach to the 4T 2T layout is shown in figure 2.8. Here the cold feed is connected upstream of the vent so pump-over is not a risk. In order to prevent air from bubbling up the cold feed it is connected to the under side of the flow pipe via a loop. Avoiding the risk of pump-over makes the pressure difference between the two connections less critical. It is a useful design if for some reason the distance between the two connections has to be longer than the recommended 150mm maximum.

Figure 2.9 shows a layout that is used with boilers that have separate connections for flow to the space heating and primary hot water circuits. It is functionally similar to the layout of fi gure 2.6. A vent chamber or trap is another variant on the 4T2T layout that aids venting and traps sediment (figure 2.10).

Disadvantages of open vented systems are the need for a cistern in the attic which adds to the installation cost and provides a potential site for legionella growth (see chapter 4). If a leak occurs the lost water is continuously replaced by water flowing through the float valve in the cistern. Cisterns are a particular problem as they are normally located above the main insulation of the house where the temperature may drop below zero. If the heating is turned off or to a low setting for a period of days in cold weather, the water in the pipe supplying high-pressure cold water to the cistern may freeze, splitting the pipe. When warmer weather comes the temperature rises, melting the ice in the pipe and releasing a torrential leak. This is prone to cause particularly destructive flooding because the leak is in the attic and soaks all ceilings, walls and rooms below it. The danger is made worse because these leaks usually occur when the house occupants are away as that is when the heating is set to off or low for several days. The leak can keep pouring for days before being detected.

Further disadvantages of open vented systems are that the water in a cistern is exposed to the atmosphere and oxygen can dissolve in it. This promotes oxidation of the internal surface of the system's components. Also because the system runs at relatively low pressure, problems such as pump cavitation and micro air leaks can arise.

Open vented systems have their advantages and remain a good option in certain circumstances, specifically where a conventional primary hot water circuit and hot water storage tank are to be used with a conventional boiler. They use low operating pressures and consequently low stress on components. The system runs at a relatively

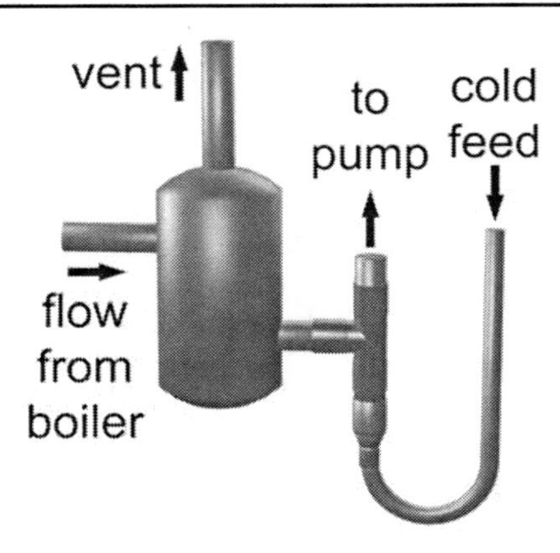

Figure 2.10: A vent chamber is wider than the flow piping so water entering it slows down. This aids separation of suspended solids which fall to the bottom of the "trap", and bubbles which rise up the vent. This is the best 4T2T system.

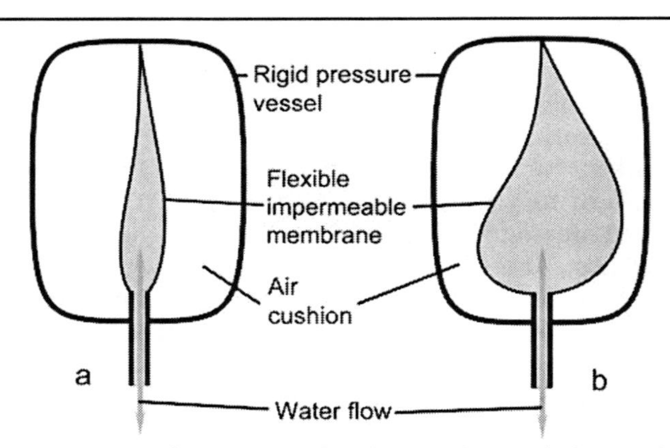

Figure 2.11: A type of expansion chamber used in sealed central heating systems. Water expansion is accommodated by water flowing into a chamber with flexible impermeable walls. Surrounding this chamber is a rigid pressure vessel and between the vessel and chamber is a space filled with gas – the gas cushion. Under normal operating conditions when the central heating system is cold, the pressure in the cushion is about 1 atmosphere and the flexible chamber is relatively narrow (a). When the system is hot, water expands and flows into the flexible chamber, compressing the air further and raising the pressure to over 2 atmospheres (b).

constant pressure and while operating normally there is minimal need for attention from the user.

Sealed systems

The last 15 years have seen increasing use of the alternative sealed systems. This is largely because of the widespread popularity of "system boilers" that containing all the necessary components for a sealed system other than the radiators. This greatly simpli es installation. Serious errors of installation causing rapid decay or malfunction are much less likely with sealed systems.

In a sealed system, thermal water expansion is accommodated with a specially design vessel that does not allow the water to come into direct contact with the atmosphere. Figure 2.11 shows how it works. This change brings several advantages. This system is easier to install as there is no need to run pipe work to the attic for a cistern and vent. It reduces corrosion because oxygen cannot dissolve in the water as there is no air/water interface. Also the risk of major water damage is avoided because there is no mains pressure piping above the attic insulation where it may freeze, and even if a leak does occur only the water contained within the system can escape as there is no oat valve top up system to automatically replace losses.

System boilers contain an expansion vessel within the boiler casing. Sealed systems are run at a pressure of 1 atmosphere when the system is cold usually rising to 2½ when hot. A pressure release valve is set at three atmospheres. As there is no automatic means of topping up the water this must be done manually via a lling loop.

Independent water and heating

Some heating installations use different boilers for the hot water and radiators. This usually arises because of the history of heating in the house. If the house originally had no boiler, no central heating, and hot water provided by an immersion heater, a boiler may have been installed to heat water because of the high cost of running an immersion heater. If at a later date it was decided to install a central heating system a separate boiler may have been used as the hot water was already catered for. This brings the advantage that breakdown of one system does not affect the other. The disadvantage is that the hot water boiler will usually be an old inefficient one. Another situation were dual systems may be used is when a very large house is being heated. Commercially available domestic boilers go up the total power of around 40 kW. This is generally enough to heat a fairly large house;

up to about ve bedrooms. For very large houses it will be inadequate and there is a gap in the market between the largest domestic boilers and the smallest ones intended for industrial or commercial use. Furthermore industrial boilers are considerably more expensive. It is often economic, and convenient, to install two domestic boilers dedicated to different parts of the house. One common solution is to use one boiler for the upstairs and one for the downstairs and to have one of these boilers also providing hot water. Among the numerous possibilities a good arrangement is to use a powerful combi boiler to heat the hot water via a thermal store system (see chapter 3) and the downstairs, and to use a smaller standard system boiler for the upstairs.

Chapter 3
Hot Water Circuits

Most central heating systems serve the two functions of space heating and heating water for the hot taps. We will take a look at the second function in detail here.

A number of hot water systems are in use, most of which use one boiler for both space heating and hot water. There are some systems with separate boilers but they are rare and usually installed because of the history heating in the building (see chapter 2).

Most installed systems using one boiler for both tasks cannot heat both radiators and hot water together. They are almost always "hot water priority". This means that when both space heat and hot water heat are called for, the boiler will first deliver heat to the water until satisfied and only then deliver space heat. The availability of mid position three port motorised valves makes it possible to deliver heat to both systems at once but this is not necessarily a good idea. When the hot water system is calling for heat, the occupants of the building may regard this as a higher priority than space heat. The power available to heat the water will be lower, and so the time taken to heat the whole tank longer if some of the boiler's output is going to the radiators.

Until recently the standard installation was the convective or gravity fed low pressure hot water storage tank. This system no longer complies with the UK building regulations for new houses. Newly built dwellings now have to have systems where the hot water is "potable" or in other words it can be drunk. Systems that comply with this requirement include directly heated water without any storage facility as provided by combination or "combi" boilers, un-vented storage tanks and thermal store systems. These latter two deliver hot water at mains pressure. They are the best performing hot water systems and will be the installations of choice in the future. Low pressure systems can be potable with modern designs and materials but cannot be convectively heated for new installations.

Gravity fed hot water cylinder

The convective system is illustrated in figure 3.1. A copper, insulated cylinder which usually contains between 60 and 245 litres is connected to a filling system. The water within the tank must not be allowed to mix with any water that has passed through the boiler. Instead it is heated indirectly by water flowing in the primary hot water circuit. This is made up of a flow pipe that brings water from the flow side of the boiler to a copper coil located in the lower half of the hot water cylinder, and a return pipe carrying it back to the return side of the boiler. The coil is a heat exchanger. Hot water from the boiler gives up its heat to the water in the tank without mixing with it when passing through the heat exchanger. There may or may not be a cock or control valve in the flow side of the circuit.

In the original design shown in figure 3.1 that was all there was to it. There are no automatic valves or motors controlling flow through the circuit. The cylinder is installed one or more floors above the boiler and convective currents drive water round the primary circuit when the pump is off. Water in the flow side on its way from the boiler to the heat exchanger is hotter so rises, water on the return side on its way back is cooler so sinks. Convective currents are driven by fairly low pressures so the pipe work of this type of primary hot water circuit

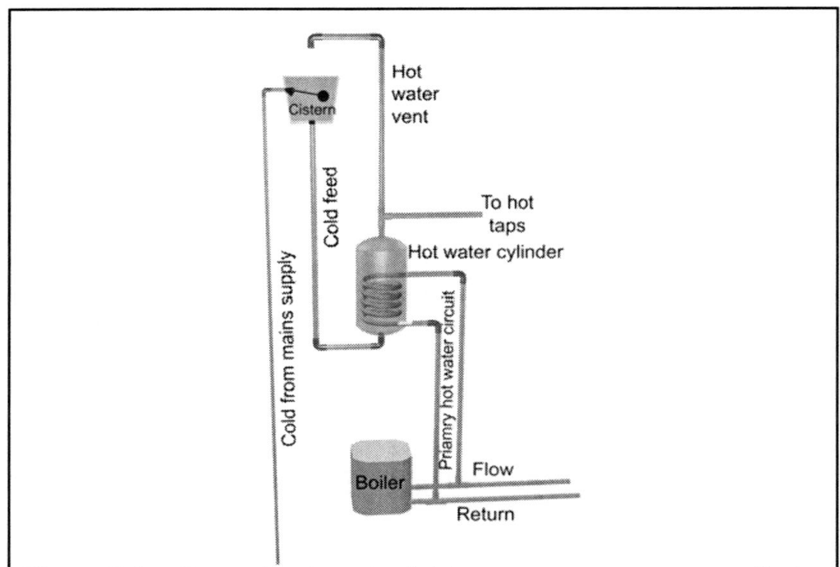

Figure 3.1: Conventional gravity fed hot water systems were usually installed over three floors. The cistern was placed in the attic to allow an adequate hot water pressure at the taps. The tank was placed on the first floor to allow enough height for convective currents to drive water around the primary hot water circuit.

must be of a substantial diameter and be as straight and free from obstructions as possible. Twenty two or 28mm rather than 15mm copper tubing is used.

The convective system is reliable and works fairly well but suffers from inefficiencies because slow flow contributes to its being unresponsive and the entire hot water cylinder ends up being fully heated whenever space heating is in use. Pressure to improve heating efficiency and reduce fossil fuel use has led to tightening of building regulations and the system has not been allowed as a primary installation since 2002 in the UK. There are still millions of such systems in use. Where feasible they should be improved with a simple and inexpensive upgrade. All that is required is to install a tank thermostat if one is not already in place, and to replace the simple flow T with a motorised 3 port valve controlled by the thermostat (wiring diagram figure 14.6)

Pumped hot water cylinder

A pumped hot water cylinder forms part of a fully pumped system. It is the same as the gravity fed system outlined above in that it consists of a circuit carrying water from the flows side of the boiler through a heat exchange in the bottom of the hot water tank back down to the return side of the boiler. The difference is that a pump drives water round the circuit rather than convective currents. Pumped systems are generally more rapidly responsive and efficient than gravity fed systems. They have the additional advantage that it is not necessary to mount the hot water tank on the floor above the boiler which may occasionally be useful.

Open vented hot water tank

Open vented hot water tanks are an integral part of the previously standard gravity fed hot water system and many pumped hot water system. They merit their own heading because they are common to these systems and to less common ones like those using solar heating. The system used for off-peak electric heating is closely related and is described below.

As the hot water flowing in the primary circuit passes through the heat exchanger it gives up its heat to the surrounding water in the cylinder which expands and becomes less dense so rises up the cylinder to float at the top above the cooler water. By continually heating the water at the bottom of the cylinder, hot water collects at the top and the hot/cold interface gradually moves down the tank. Hot water is drawn off the top of the cylinder so that it is not necessary to wait until the entire cylinder is hot before hot water can be used. The system uses hot water at a controlled, fairly low pressure so cannot be connected directly to the mains water supply. Instead it is fed cold water from a float valve served cistern in the attic. The layout is shown in figure 3.1.

Hot water for the taps is drawn off from the highest point in the cylinder. Provision must also be made to allow any gas that collects at the top of the cylinder to escape. These dual requirements are met by the layout shown in figure 3.1 where a vent pipe rises out of the top of the cylinder to a level above the cistern in the attic and the hot water is drawn off from a T connector in this pipe a short distance above the top of the cylinder. In order to prevent water from pouring out of the vent pipe (called siphoning) it must be higher than the level of water in the cistern. Not only must it be higher but it must be higher by at least 4% of the height of the level of water in the cistern above the bottom of the hot water cylinder, plus a safety margin of 150 mm. To understand the reason for this 4% of extra height think of the U tube formed by the cistern, cold feed, cylinder and vent. On the vent side of this U the water is hot whereas on the cold feed side it is cold. This means that the hydrostatic pressure at the bottom of the U can support a higher column of the hotter, less dense water on the vent side than on the cold feed side. Consequently the less dense hot water will rise up further in the pipe than the level of water in the cistern. Figure 2.3 shows the maximum density range to be from 1 at 4 °C to 0.96 at 100 °C or 4%. This high temperature should never be reached unless there is a malfunction but it is better to design for that.

Strictly the height of the vent pipe refers to the height of the opening of the pipe rather than the high point of the bend though in practice with 22mm or larger pipe it is unlikely that siphoning will occur. To make sure a T connected can be put in the copper pipe facing upwards at the high point covered by a grill to exclude insects as shown in figure 3.2. This is handy if the height above the cistern is so great that the water may splash out over the sides or there is not enough room to install the full height directly over the cistern. In high Victorian three-storey house this can be a problem.

Cylinders for vented systems come in three grades according to the pressure they are designed to withstand. This pressure is governed by how high the cistern is above the bottom of the tank. The height of water that equates to one atmosphere pressure is 10m. Grade 1 tanks are designed for a maximum cistern height of 25m. Grade 2 for a maximum of 15m, and grade 3 for a maximum of 10m. As you might expect the greater the pressure handling capacity, the higher the cost of the tank. Much the commonest tank to install is a grade 3. High-grade tanks are only necessary when there are more than two floors between the cylinder and cistern.

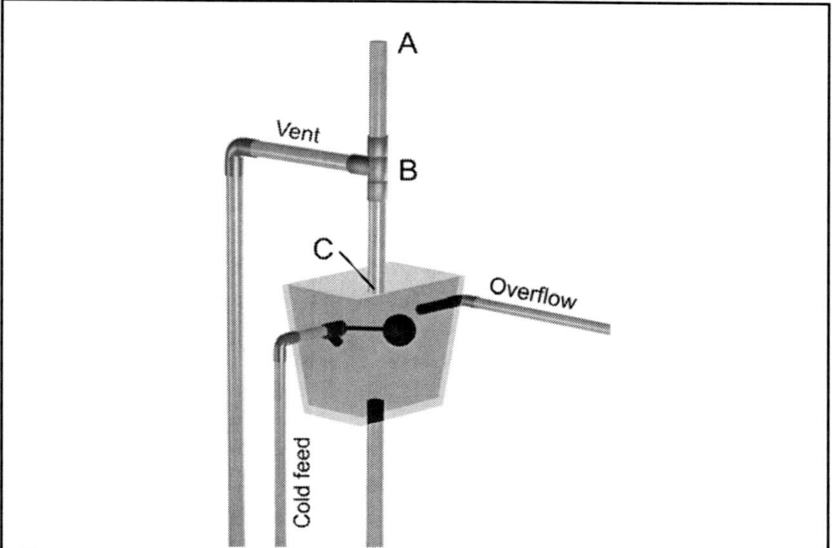

Figure 3.2: An arrangement that can be useful in tall houses where the height of the hot water vent above the cistern is substantial. A T connector is used with the upper end open to the air (A) and covered with a corrosion resistant mesh with a hole size not greater than 0.65mm to exclude insects. This prevents the loop over of the vent from acting as a siphon. The effective height over the cistern water level is now to B rather than C. The down pipe from B to C need not be vertical.

The pressure of water at a particular tap and consequently the flow rate from that tap is determined by the height of the cistern above the tap. The connections between the cistern and the hot water tank and between the top of hot water tank and the run off to the various hot taps are with 22 mm diameter pipe. 15mm pipe is used for the run-offs.

The disadvantages of vented systems are that many older hot water tanks are poorly insulated thus leading to heat losses and increased cost. Another problem is that the pressure of hot water is limited by the height of the cistern in the attic. This will usually be substantially lower than the pressure of the rising main. This leads to difficulty in balancing showers. Often the most irksome disadvantage of the system is that the amount of hot water is limited. It takes a considerable time to heat the whole tank of hot water and it is usually enough to supply only one large bath. This leads to the major drawback that you cannot draw several baths in sequence.

Primatic cylinders

"Primatic" is a trade name for what is technically known as a single feed cylinder. It is a system which fills the central heating from the hot water storage tank in the attic so avoids the need for a separate central heating expansion cistern. Figure 3.3 shows how it works. The system of domes and vents causes water to spill over from the hot water circuit into the central heating circuit when the level in the central heating circuit falls. The separation of hot tap water from central heating water is effected by the air trapped under these domes. The disadvantage of the system is that if some malfunction such as gas collection in radiators causes the level of water in the central heating system to rise it can spill over into the secondary hot water circuit so the separation between central heating water and hot tap water is not as reliable as in other systems. For this reason inhibitors should not be used. The system was briefly popular after its introduction in the mid 1970s but is now obsolete and should be replaced where possible. The way to tell that the system is primatic is that there is a hot water storage cistern in the attic but no central heating expansion cistern though the central heating is open vented and has no sealed expansion tank.

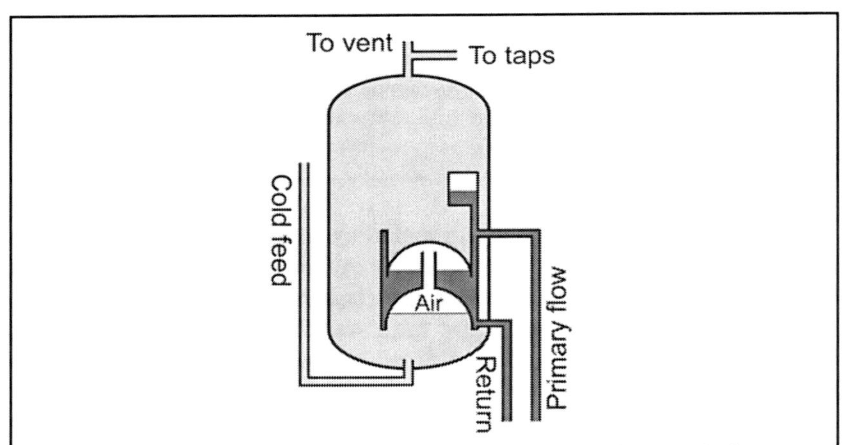

Figure 3.3: A primatic or single feed hot water cylinder. Both the space heating and hot water are toped up by a single cold feed. This is directly connected to the hot water circuit (light grey) as in a conventional system. If the water level in the space heating circuit (dark grey) is low the water level trapped below the air under the lower dome rises up the pipe between the domes until it spills over the top into the space heating circuit so topping it up. The problem is that if a fault such as gas accumulation in radiators makes the water level in the space heating circuit rise, space heating water may spill into the hot water circuit. Because of this risk inhibitors (chapter 6), which are toxic, should not be used. The system is now obsolete.

Electric heating

Hot water cylinders almost always contain an immersion heater. This is installed as a backup against boiler failure. The problem is that the electricity is likely to be roughly twice as expensive as the oil or gas used to fire the boiler to heat the same quantity of water. Electricity is used as a primary method of heating with cheaper off peak power. Cylinders designed for this kind of system have the simplification of requiring no coiled pipe heat exchanger. They are called direct cylinders (figure 3.4) because they are heated directly rather than via a primary hot water circuit. They are cheaper to buy and easier to install. The off peak electricity is typically available during the hours after midnight to 5 am so for the system to perform satisfactorily requires that the hot tank is well insulated so that it can remain hot from the early hours of the morning until the following evening when most hot water is used, and is of a substantial size so the entire household daily requirement for hot water can be contained within it. These demands are eased when off peak electricity is also available in the early afternoon as it often is.

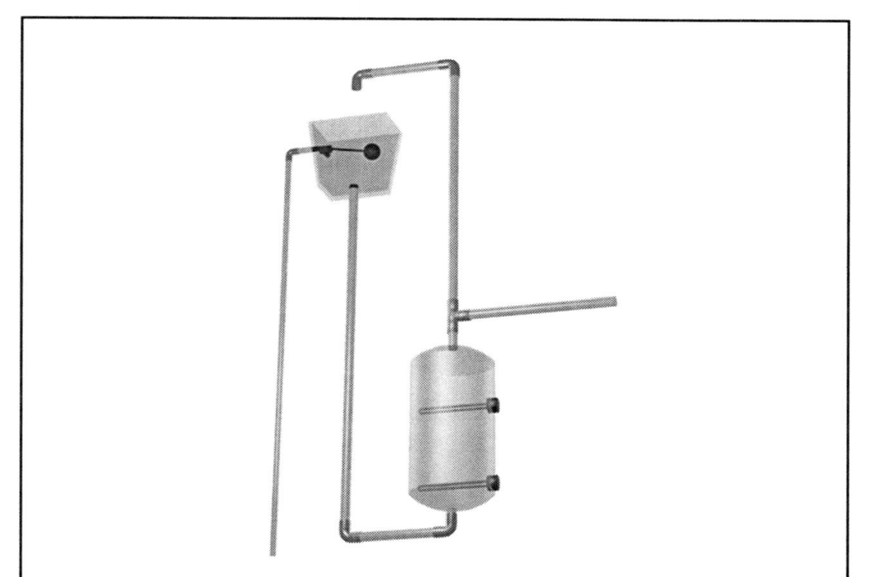

Figure 3.4: Electric hot water systems work in the same way as conventional gravity fed systems except there is no primary hot water circuit. Instead they have two immersion heaters. The lower one is intended for use with cheaper off-peak electricity during the night and heats up the whole cylinder. The upper one is for top up heating. It uses the more expensive standard rate electricity and only heats up the upper third of the tank.

Immersion heaters

Electric heating is done with immersion heaters that are "immersed" in the water in the tank and heat it directly as does the element of an electric kettle. The usual immersion heaters fitted are 275mm 3 kW types with 2¼" BSP boss threads. Off peak electric systems use 2 immersion heaters, one located about 150 mm above the base of the cylinder and the other about two-thirds of the way up. The lower one is used to heat the water with cheap electricity at night and heats up the whole tank. The upper one is used for top up heating during the day if there is extra demand, only heats the top third of the tank and uses more expensive daytime electricity to do so. Both emersion heaters are usually rated at 3kW. They are connected to the mains electricity supply with a three core heat resistant flex rated at 20 amps. All emersion heaters are wired to the mains supply via a double pole 20 amp switch with a neon indicator. If off peak electricity is in use the lower immersion heater will be controlled by a time switch that coincides with the off peak periods.

Cylinders designed to be indirectly heated still have a 2¼" boss intended for a back up immersion heater. It is not essential to fit an immersion heater in this situation and the boss can be blanked off but this is not recommended. Considering the inconvenience of losing both central heating and hot water in the event of a boiler failure, and the likelihood of such a failure, the small additional cost of installing an immersion it is well worth it! Unlike direct cylinders, in indirect cylinders the emersion heater is screwed into the top of the cylinder and sticks downwards into the water rather than being screwed into the side and sticking sideways. As cylinders come in a number of different lengths, different lengths of immersion heater are made to fit them. In general the immersion heater should be the longest size which is less than three-quarters of the length and cylinder. In most cases a 695 mm one is used.

Long life emersion heaters with titanium elements are available for a slightly increased cost. They are recommended in hard water areas.

Immersion heaters have built-in thermostats that can be adjusted for water temperatures between 10 and 80°C. They are factory preset at 60°C. Some have an additional independent over temperature cut out that switches the heater off should the regular thermostat malfunction and the water temperate reaches 98°C.

It is possible to replace the thermostat in an immersion heater without removing the heater from the cylinder and having to drain down the system. This occasionally may be worth doing but in general when the thermostat fails it is a good idea to replace the whole heater.

Un-vented cylinders

Un-vented cylinders are also indirectly heated but they differ from conventional cylinders in that no vent pipe or storage cistern is used. Rather they are connected directly to the mains cold water supply via a pressure regulator that maintains a pressure in the tank of three atmospheres. This means that they operate near mains cold water pressure which is considerably higher than that found in vented systems. This gives the advantage of high hot water flow rates. They are available in capacities ranging from 70 to 300 l. The expansion of the water as it is a heated is accommodated by a gas filled expansion vessel which may be incorporated into the cylinder or installed outside. They are proofed to much higher pressures than ordinarily tanks at between 12 and 15 atmospheres. They also incorporate pressure release valves that are connected to pipe-work that diverts water to the outside of the building in the event of an excess pressure arising within the tank. The potential risks associated with storing large volumes of hot water under high-pressure has led to extensive regulations controlling the manufacture and installation of these tanks. In the UK installation must be carried out by a "competent" installer for any vessel containing more than 15 l of hot water at mains pressure. What is meant by "competent" is defined in the UK building regulations approved document G3 and included any person registered with a suitable UK professional body.

Combi boiler systems

Combination, usually known as "combi", boilers are discussed in greater detail in chapter 11. The principal difference between the operation of a combi boiler and of a conventional boiler is in the way they handle the hot water system. The combi boiler heats water directly as it is required. When a hot tap is turned on, a pressure sensor in the hot water piping within the boiler senses a drop in pressure. This triggers the control system to operate electrically activated valves to divert water from the rising main through a heat exchanger and into the hot water circuit, and to fire the boiler. The rate at which hot water can be delivered is limited by the power of the boiler. Because this limiting factor is a major mark of the convenience of combi boilers, they tend to be more powerful than their conventional counterparts. Table 3.1 gives the hot water flow rates possible with different combination boiler powers.

Table 3.1. Powers and hot water flow rates from combi boilers at temperature rises or 35 and 55 °			
Power kW	Equivalent Power BTU	Flow l/min with a 35 °C temperature rise	Flow l/min with a 55 °C temperature rise
22	75000	9.0	5.7
25	85000	10.2	6.5
30	102000	12.2	7.8
35	119000	14.3	9.1
40	137000	16.3	10.4

The advantages of the combination boiler system are that a limitless supply of hot water is available as is it is heated continuously; also there is no cylinder of hot water and so no associated installation costs or ongoing heat loss.

The disadvantages of the system are that combi boiler technology is highly complex, and earlier models were prone to failure. The hot water flow rate is significantly limited particularly on smaller combi boilers and even larger boilers suitable for heating large houses are not able adequately to supply more than one or two hot taps at once. A further disadvantage is that it takes some time for the water temperature to stabilise. The temperature goes through phases of cold and scorching hot before finally stabilising to the preset temperature. Although combi boilers take their water directly off the rising main they use a pressure reducing valve before passing it through the heat exchanger. The hot water pressure is therefore considerably lower than the rising main pressure, as is the case for the vented hot water tank system.

The advantage of having a limitless supply of hot water is so great that notwithstanding these other problems, combination boilers have become very popular. Many of the problems associated with them can be solved by using specific types of thermal store water systems as described below.

Thermal store hot water systems

Thermal store hot water systems are also referred to as mass storage or water jacketed heater systems. There are several different arrangements for use with combination boilers, conventional boilers or other heating appliances. The principal they all have in common is that a large volume of water is kept hot; not to act as a hot water store but rather to act as a heat store. Water is the ideal material for this because despite its relatively low density it still has a very high volumetric heat capacity.

Thermal store systems bring a number of advantages. Firstly hot water is available at mains pressure and this makes it easy to balance showers. It is possible to supply hot water at high flow rates for a limited period of time allowing the best performing showers possible and rapid bath filling. The system can be used with a conventional or combi boiler. A surprisingly valuable advantage is that the water in the tank can be kept considerably hotter than in other systems. This is partly because of the very high performance of the installation of thermal store tanks but mainly because there is very little change over of water in the tank. This means there is no continuing supply of calcium carbonate to produce lime scale. High temperatures can be maintained as once all the lime from the water has precipitated, no more will arrive. The tank can readily be kept at 70 °C or more. This is too hot to pipe straight into the hot water taps especially a mains pressure! Instead a thermostatic mixing valve is used which combines water coming from the heat exchanger of the thermal store tank with cold water in appropriate amounts to provide out flowing water at 60°C or what ever temperature is chosen. In turn this means that there is more hot water available at 60°C than has to be stored in the tank at 70°C or above. This strategy coupled with the large size of mass storage tanks means that their heat capacity is very generous. They can frequently fill several baths before they become cool. This means that a smaller boiler can be used than would otherwise be necessary. As you might expect they are a very good choice for off-peak cheap-rate electricity powered systems.

A further advantage of keeping the tank hotter than the hot water and using a thermostatic mixing valve is that the temperature of water coming from the hot tap does not fall at all until the temperature of water in the tank has declined substantially from 70 to 60°C. With this much capacity it is possible to operate the boiler for prolonged periods with fewer on-off cycles, enhancing reliability and efficiency.

The essential system requirements are a large well insulated tank to contain the water with some means to heat it and some means to extract the heat from it. Additional parts are necessary to keep the tank full of water. These consist of a cistern with a float valve as is the case with a gravity fed hot water tank but as the pressure of water in the store is not critical it can be much lower than it is for a conventional system. This means that there is no need for a cistern in the attic, avoiding the risk of freezing, burst pipes and flooding. It is usually mounted immediately above the thermal store.

There are a variety of products on the market that achieve these aims in differing ways. One commonly encountered arrangement is shown in figure 3.5. In this example the water in the mass storage tank is heated indirectly by hot water from the boiler passing through a heat exchanger at the bottom of the tank as is found in a conventional indirect hot water cylinder. Hot water for the taps is taken directly from the cold water mains supply and heated by passing it through a second heat exchanger external to the thermal store tank. This arrangement involves a fairly complex control and pumping system but has the advantage that lime bearing water from the mains is not heated to as high a temperature as the water in the tank. This avoids lime scale deposition inside the piping carrying mains water.

The system shown in figure 3.6 uses a second coil located inside the top of the tank as the second heat exchanger. These are longer than the primary heat exchanger and quite often finned to allow faster transfer of heat. The disadvantage of this system is the potential for lime scale to form on the inside of the second heat exchanger because mains water can be heated to the same temperature as the water in the tank which may be over 60°C.

Thermal store tanks may also contain immersion heaters as backup against boiler failure and there may be additional heat exchangers. For example there may be one to deliver heat to the tank from solar heating panels and another to extract heat from the tank for space heating. The water in the hot water store can be treated as would the heating water in a central heating circuit and passed directly through the boiler avoiding the primary heat exchanger.

A powerful combi boiler can reheat the hot water store sufficiently quickly that even though the highest flow rates cannot be maintained for long, water can be reheated adequately to run sequential baths even with fairly small thermal store tanks. Used in this way thermal store systems solve the problem of swings in temperature just after a hot tap is turned on that combi boilers are prone to.

With a thermal storage system the hot water coming from the taps has not been stored. There is no risk of legionella infection and the water can be drunk.

The disadvantages of the system are that it occupies quite a lot of space as does the hot water cylinder system. These days the problem of heat loss from the surface of the hot water tank is minimised because they are supplied with high-quality insulation that far outperforms the earlier tie on jackets. The system is also more costly to install thanks to the size and complexity of the storage tank. When compared to the price of installing a heating system this additional cost is fairly modest and will generally be found the well worth while given the far superior performance afforded.

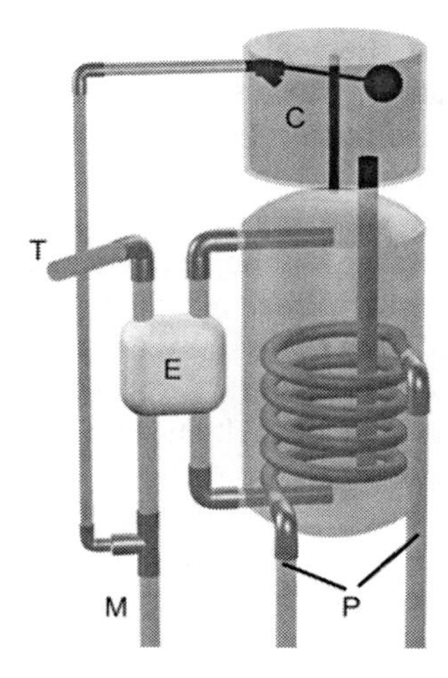

Figure 3.5: One form of thermal store hot water system. This one uses a heat exchanger (E) external to the water tank. Hot water flows through the heat exchanger from the tank and gives up its heat to water flowing in from the cold mains (M) before going to the hot taps (T). There is no requirement for high pressure in the tank so the cistern (C) is mounted directly above it. Here the thermal store is heated by a primary hot water circuit (P) but it could equally be heated by immersion heaters or use a system where the water contained within the thermal store is pumped directly through the boiler.

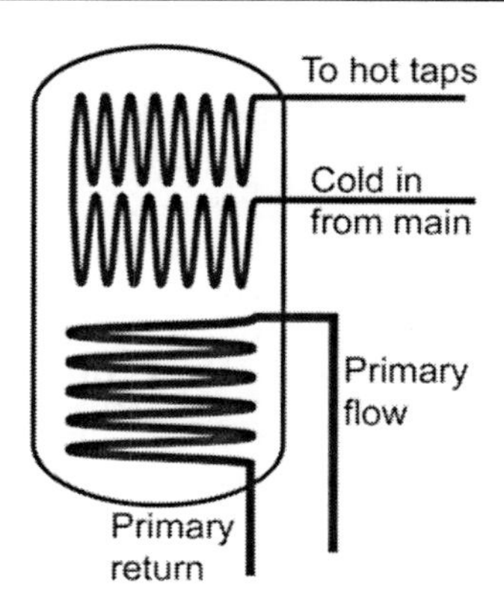

Figure 3.6: This is another type of thermal store system where the secondary heat exchanger is a pipe coil within the upper half of the cylinder. A limitation of this system is that lime scale deposition inside the secondary heat exchanger is a risk if the temperature of the thermal store is over 60°C.

Mains pressure hot water is becoming increasingly popular and the two technologies of thermal stores and un-vented tanks are in competition for this market. Both systems have their advantages. Un-vented tanks are simple in principle and so tend to be reliable. Thermal stores do not have to be installed or maintained by M3 registered persons because they contain less than 15 l of hot water at mains pressure. Also they provide more hot water for their size because of the ability to run them hotter. The constant inflow of water to an un-vented cylinder would cause build-up of lime scale if it were run at these temperatures.

The costs of thermal stores and un-vented cylincers are comparable to the cost of the boilers that heat them with un-vented cylinders being marginally cheaper.

A point that may be important is that as thermal store tanks operate at low pressure they do not have to be round. Some companies produce rectangular tanks that fit neatly into many of the rectangular cupboards and spaces where hot water systems tend to be installed. This may allow a larger hot water capacity to be installed in a particular location than would be possible with a cylindrical tank.

Hot water temperature control

"Control" is possibly not the right word to use here as regulation of the temperature of hot water coming out of hot taps is one of the most unsatisfactory aspects of modern domestic heating systems. At the best of times a hot tap must be opened and a considerable amount of cold water run out of it before it "runs hot". Combi boilers have a particular sin of taking some time before the temperature stabilises. During this time water coming out of the boiler alternates between stone cold and very hot and the temperature changes can be quite sudden. The ideal solution to these problems has yet to become available but they can at least be minimised by careful planning of the hot water system.

Table 3.2. Comfortable temperature ranges for various hot water applications

Application	Temperature range °C
Baths	41-44
Showers	42-45
Hand washing basins	38-41
Bidets	36-38
Kitchen washing-up	45-50

Table 3.2 shows the comfortable temperature ranges for various applications. The ranges are quite narrow. To get the temperature within these ranges the water coming from the hot tap has to be hotter so that it can be diluted with cold water to get the desired temperature. As the hot water is stored remote from the taps it must be conveyed to them by pipes and during its journey to the tap it will inevitably lose some heat. For this reason the hot water tank is kept at a temperature above that desired for water coming out of the taps. Unfortunately the amount of heat lost by the water on his way to the taps is rather unpredictable. Hot water is stored at 60°C which is hot enough to cause scalds. This temperature is chosen mainly to avoid the risk of legionella growth (see chapter 5) and to minimise lime scale formation rather than for reasons of comfort. It is distributed at a nominal temperature of 55°C with the objective of hot water coming out of hot taps at a temperature of 40 to 50°C. In reality the temperature of hot water coming out taps can be anything from under 10 to 60°C. It should never be over 60°C. In practice a modern installation with the hot tank kept at 60°C will often be too hot at the taps. The natural reaction to this by the users will be to turn down the tank thermostat. This is not encouraged because of the legionella risk. Thermostatic mixing valves for use with showers are familiar. Similar valves are available for use with hot water circuits and taps. They have inputs from the hot and cold pipes and mix the proportions in order to arrive at the desired temperature, which is usually set to be between 40 and 50°C.

We have all grown up with vagaries of hot water systems and have learned caution so that most of us can operate a hot tap safely without supervision. This cannot be assumed if a hot tap is installed in a location where it will be used by persons of diminished responsibility, in which case a thermostatic control system should be used.

Cylinder thermostat

Cylinder thermostats, sometimes known as tank stats, are heat sensitive switches that can be attached to the outer surface of the hot water cylinder. Their operation is described in more detail in chapter 12. To detect the cylinder temperature accurately requires a good thermal contact between the heat plate of the thermostat and the cylinder. There must be no insulation between the two. In bare cylinders they are simply placed on the surface. With jacket lagged cylinders they are located underneath the jacket. With foam insulated cylinders a small part of the foam must be cut away exposing the bare copper and the copper cleaned of all insulation with wire wool or some other abrasive to accommodate the hotplate of the thermostat. The thermostat is usually held in place by an elastic strap that goes round the cylinder. Net curtain wire is commonly used for this.

The length is joined by two hooks or a hook an eye. The join should be at least 50mm away from the thermostat.

There is no universal consensus on how far up the cylinder the thermostat should be located. As it is intended to turn of the heat supply when the cylinder is full of hot water it should be located fairly near the bottom. Between ¼ and 1/3 of the way up from the base is usual.

A confusing aspect of cylinder thermostats is that they are not simple two terminal switches but three terminal switches. This means that rather than switching one connection on or off at the set temperature the thermostat switches a common terminal between two others called normally open (NO - normally in this case meaning when cold) or 1 and normally closed (NC) or 2 (see figure 14.2). This gives ample opportunity to wire them up wrongly! There are two possible faults. If wired from pin 1 to pin 2 they are always off and the boiler will not heat the cylinder. If wired to the common and normally open pin the malfunction is more bizarre. The cylinder will not heat from cold. The usual immediate response to that is to turn on the emersion heater. That heats up the cylinder and depending on the relationship between the set temperatures of the immersion heater and thermostat, the thermostat may switch on when the cylinder is hot and stay on irrespective off how hot it gets which will then be determined by the boiler thermostat.

Cylinder thermostats are now obligatory on new systems. They allow much better control of cylinder temperature which makes the system more energy-efficient, cost-effective, safer, and less prone to lime scale build up.

Insulation

The regulations in the UK regarding what kind of insulation must be used in hot water systems have been tightened considerably in recent years. The earliest system of a bare cylinder or a cylinder insulated by a removable jacket is no longer allowed for a primary installation. Newly installed hot water cylinders must be pre-insulated. Tie on insulation jackets are still available for use with existing installations.

A further recently introduced requirement is that all pipes both hot and cold must be insulated within 1 m of the cylinder.

Beyond one metre from the cylinder regulations do not require insulation but long uninsulated pipe runs to taps significantly cool water on its way. It is recommended that long runs are insulated. The length beyond which installation should be used is variable but it relates more to the volume of water held in the pipe than to its overall length so the smaller the diameter the longer run without installation. In pipes of 28 mm a run of anything over 3 m should be insulated. This increases the 12 m for 15 to 22 mm pipes and 20 m for smaller pipes.

CHAPTER 4
Health and Safety

Legionella

In 1976 in Philadelphia, Pennsylvania, a mysterious outbreak of a previously unknown pneumonia occurred. Over a short space of time a total of 221 cases were noted by Pennsylvania public health officials. The pneumonia was a particularly severe one with more than 60% of cases being hospitalised and 34 deaths. It was quickly realise that all cases where members of the American Legion; so-called legionaries. The disease became known as legionnaires' disease. Furthermore all of the cases had attended the annual Pennsylvania State legionnaires' convention held in a Philadelphia hotel that year. Public health officials naturally suspected that there had been an outbreak of an infection that spread from person to person. What they eventually discovered was quite different. The disease was not transmitted from person to person but was acquired by inhaling an aerosol of infected water in this case from a cooling tower by the hotel. The cause was a previously unrecognised bacterium. It was named legionella. The outbreak caused considerable interest and tissue samples from similar outbreaks that had been collected over the previous 50 years were re-examined. Sure enough in many cases the legionella bacteria were found. It was not a new disease that appeared in 1976 but that was when it was first recognised and identified.

Legionella is important for central heating and water installations because of its serious and life-threatening nature and because it is specifically contracted from water handling appliances.

Legionella is ubiquitous in natural ground and surface water. It tends to grow in biofilms[1] and on the surface of lakes, streams and rivers. It can grow in association with certain larger single celled organisms like amoebae that protect it from the disinfectant

[1] Biofilm is a brown film of algae and bacteria that forms on surfaces submerged below stagnant fresh water for long periods.

effects of many chemicals. This enables it to survive water treatment processing and it can be recovered in small numbers from domestic freshwater supplies. It becomes infectious if allowed to multiply raising the dose in atomised water by thousands of times. Control of legionella thus focuses not on elimination but on preventing multiplication.

Legionella is a bacterium which grows and divides in stationary water such a blind pipe ends or storage tanks between 20 and 50°C with a particular risk between 35 and 46°C and pH between 5 and 8.5. It is killed by temperatures of 60°C or above which in large part is why this is chosen for hot water storage temperatures. Algae, biofilms, and deposits that come under the general description of poor hygiene promote its growth. Instrumental in avoiding legionella risk is to avoid these circumstances as far as possible. In central heating systems where a large volume of water is kept at approximately the correct temperature for legionella, much of it in stagnant corners, the risk must be carefully managed.

Various precautions are necessary to prevent legionella growing in water systems:

Arrange that the cold water system remains below 20°C. Avoid long runs were cold pipes travel next to hot pipes. A particular area of risk is cold water storage cisterns which should be situated in cool places and protected as far as possible from sunlight and have lids fitted to exclude insects. Cisterns should be an appropriate size for the installation so that they regularly drain and refill in normal operation without leaving a large volume of stagnant water where biofilm can accumulate. Equally storage cisterns that are too small should not be installed with numerous ones linked together. They should also be equipped with back flow prevention valves. Storage cisterns should be checked annually and if any signs of biofilm are found they should be treated chemically with chlorine or hydrogen peroxide products.

Once legionella has developed to dangerous levels, in order to be transmitted to susceptible people it must be carried in the air as an aerosol. In the domestic environment this is most likely to happen with the hot water system. Taps, showers and spraying devices are all potential sources of atomised water. In order to avoid these being involved in the transmission of legionella it is important that the central heating water be kept strictly isolated from the hot water with the use of a primary hot water circuit. There is a still a risk if the heat exchanger coil in a hot water tank should become corroded and leak.

Carbon Monoxide

Carbon monoxide is a particular problem because it is poisonous to man. It has the property of binding to haemoglobin[2] 200 – 250 times as tightly as oxygen. Inhaled carbon monoxide occupies and hence blocks the oxygen binding sites, stopping blood carrying oxygen, a situation which may be fatal. To make matters worse carbon monoxide is colourless and odourless. It is produced when combustion occurs with too little air and poisoning incidents usually arise from poorly maintained and ventilated heating appliances.

A considerable amount of legislation exists to minimise carbon monoxide poisoning. It is one important reason for the legal requirement that heating appliances and their ventilation systems are properly maintained. Ventilators in the same room as fuel burning appliances should not be covered and should be designed so as not to cause cold drafts or allow noise entry to minimise the risk of their being deliberately blocked. There are various devices on the market which detect carbon monoxide in the atmosphere and they are invaluable aids to safety in areas using gas appliances. As a general rule, if an appliance which is designed to work with a blue flame and no soot is actually producing a yellow flame or soot, then carbon monoxide production is likely and the appliance should not be used pending further investigation. It is not a trivial matter. Approximately 50 people in the UK and 800 in the US die from carbon monoxide poisoning every year and many more are brain damaged.

Electrical safety

In the UK we use an AC mains electricity supply of 240 V RMS[3] at 50 Hz. There are many sound engineering reasons why these figures have been chosen but they unfortunately make the mains supply dangerous. This is partly because of the comparatively high voltage used and partly because 50 Hz is about the best frequency for provoking a fatal disorder of the heart rhythm known as ventricular fibrillation. Because of the

[2] Haemoglobin is the protein in the blood which carries oxygen from the lungs to the tissues. It is red when bound to oxygen and dark blue/black when not. That is why someone who is asphyxiated turns blue. Carbon monoxide binds to haemoglobin in the same place as oxygen, blocks oxygen binding and turns it bright red. This leads to the characteristic flushed bright pink colour of a carbon monoxide poisoning victim.

[3] A complication of alternating current (AC) electricity supplies is that there are a number of ways in which the magnitude of an alternating current – or voltage - can be expressed. These include peak to peak voltage, peak to mean voltage and root mean square (RMS) voltage. RMS is used because it allows the simple Ohm's law formulas of direct current calculations to be applied to AC.

dangerous nature of our electricity supply, elaborate and carefully designed safety precautions are necessary. Among the variety of possible safety systems, we in the UK have adopted earthing in common with most of the rest of the world. In general this is a good system but it has one weakness which is particularly pertinent with central heating systems. The weakness is that if the earthing fails, protection is completely lost. If the earthing of a central heating system fails and any part of the system becomes connected to the mains live either because of an internal failure or because of failure of insulation of a cable in contact with a bare copper pipe (a common situation underneath floorboards etc) touching any part of the system including radiators, taps, stainless steel sinks, or even getting into the bath, is liable to lead to instant death! That situation is so grievous that several layers of protection, each carefully thought out and meticulously installed, are used.

System earthing

It is essential that a central heating system is earthed. Not only that but it must be earthed in several places so that if electrical conductivity between parts of the system is lost, each individual part will retain an earth link. This multiplicity of earths also provides protection against the failure of individual earth links. An earth connection is usually provided by the electricity company or may be provided by a copper electrode buried in the ground under the house. In either event great care is taken by the company or installer to make sure that the earth connection provided in the house is as reliable as possible. It is the responsibility of persons installing central heating systems to install a link or links to this central house earthling point which are intended to be not only sound at the time of installation but also proof against future decay, mishaps, and incompetent meddling. A common practice is to run a 10mm^2 copper earth cable from the main house earth point to a secondary earth block for the central heating system. Though not legally required it is good practice to run two links via different routes to the earth block. Ten mm^2 earth links are then taken to each of the following: the fuel pipe to the boiler (gas or oil), the flow pipe from the boiler, the return pipe to the boiler, the cold water rising main, and the hot water system (either the hot water pipe coming out of a combi boiler or some point on the secondary hot water circuit in a hot water cylinder system).

The increasing use of plastic pipes and non metallic methods of joining metal components such as with PTFE tape has increased the chance of sections of the circuit being electrically isolated. Current regulations are strict about ensuring all parts have sound electrical bonding to earth – known as equipotential bonding. Any part of the circuit that is potentially isolated from the rest must have a reliable bonded rout to earth installed. The laws for large installations are complex and this is one area where it would be worth consulting an expert to check an installation if you are not familiar with earth bonding practices.

It is salutary to remember that in the past the cold water supply to a house was delivered via a metal pipe which ran underground and fortuitously provided a very effective earth connection to backup that provided by the electricity company. Nowadays these pipes are plastic and this additional protection has been lost.

RCD protection

A residual current device or RCD is an instrument that compares the current flowing in the live and neutral wires to an electric circuit. These two currants should be nearly identical. If they are not it means that current is leaking to earth, usually from the live wire. This indicates a fault. If somebody receives an electric shock it is almost always because current is being conducted through their body from the live to earth. When an RCD detects and imbalance of more that 30 milliamps in the live and neutral currents, it switches off the circuit. Thirty mA is chosen because it is a level at which ventricular fibrillation is not provoked, i.e. a 30 mA shock is not lethal. If an RCD is used in the mains supply circuit it prevents live-to-earth lethal electric shocks from occurring. This allows an additional layer of protection which is independent from earthing. It is now a legal requirement that RCD protection is provided whenever a house is wired or rewired. This rule is still fairly recent and many houses do not have RCD protection. The addition of a central heating system does not require that RCD protection is provided but if it is not installed already, it is strongly recommended that it be installed at the same time. RCD protection of the circuit supplying electricity to the central heating itself will certainly help but is not the whole story. A central heating system involves hot bare copper pipes that run all over the house and there is substantial risk that it will pick up live voltage from accidental mishaps involving any of numerous circuits, in which case shocks would not be prevented by RCD protection of the central heating system only. The safe solution is to install central RCD protection of the whole house. This should be done by a competent electrician. There is a complication in that smoke alarms must be on a circuit that is not RCD protected! RCD protection is a great aid to safety but cannot give full protection alone and does not absolve installers from taking due care over earthing.

Wherever possible switch off the power before working on an electric circuit. It may be necessary to have the power switched on while testing a circuit and it makes sense whenever possible to have the circuit RCD protected. It is likely that part of the toolkit for central heating work will include an extension lead. It is recommended that you purchase an RCD protected extension lead or replace the plug on your extension lead with a plug that incorporates an RCD. This will provide you with a transportable protected mains outlet to which you can connect circuits that must be on while you test them.

Working practices

Much of safe working practice is a combination of common sense and using appropriate equipment. If you are going to use a blow lamp for soldering pipes make sure you are equipped with a series of fireproof panels or heat resistant cloth pieces to protect whatever is behind the pipe from the flame. Pipe work is often located under floor boards or behind panels where tinder dry laths or shavings are easily ignited. Have wet rags and water handy!

If you are going to have to gain access to roof spaces without lighting, equip your self with a head mounted light to free your hands. Suitable models of head mounted battery powered high luminescence light emitting diode lights are widely available. Outdoor pursuits shops are a good source as they are popular for caving. Use eye protection where appropriate and try not to touch too many soldered joints before they are cool!

A recurring problem for central heating work is knee protection. So much of the work of installing central heating system is done at or below floor level that much of the installation time must be spent on hands and knees. This becomes extremely uncomfortable after an hour or two without knee pads. They are one area in particular where quality counts and I personally think it worthwhile investing in a high-quality pair before starting major central heating work.

CHAPTER 5
Radiators

Choice of radiators is governed by three issues: heat output, aesthetics, and cost. The difficulty with calculating the necessary heat output is making an accurate estimate of the amount of heat that a room requires. This is discussed further in chapter 8. Once the heat requirement has been arrived at it is a simple matter to select radiators whose total heat output adds up to that required. Generally there is a fairly consistent relationship between heat output and cost of radiators of a particular type with an approximately constant price per kilowatt. The most commonly installed radiators are the familiar pressed steel type and these are much the cheapest. Recent years have seen a marked increase in the number of more aesthetically pleasing designs on the market though these come at a premium cost.

During manufacture, radiators are pressure tested to 10 atmospheres. Their maximum working pressure is usually around 8 atmospheres but most domestic central heating systems use a much lower pressures of 0 - 3 atmospheres. The operating temperature varies between about 55 and 82 °C. The maximum temperature reached by radiators is governed by the output temperature of water from the boiler. This will generally be between 65 and 82°C, lower for condensing boilers. By the time the water gets to the radiators it will be somewhat cooler.

The term “radiator” is something of a misnomer because around 80% of the heat output is not radiant heat but convective. This is relevant to the effect of using different paints on a radiator. The type of paint used affects the radiant heat mainly rather than the convective. It is only the top coat of paint that makes any difference. Metallic paint in particular can reduce heat output of a radiator by up to 15% but in practice this effect is negligible because most radiators are sized generously for the rooms they are in and are nowadays thermostatically controlled so the only effect of slightly diminishing the heat output is to compensate by running the radiator slightly hotter.

Pressed steel radiators

By far the commonest type of radiator to be installed these days is a pressed steel model. They have a number of advantages over other types. They are considerably cheaper than any other option. In general for a given physical size of radiator the highest heat output will be from one of this type. They are also available in a wide variety of sizes with varying heat output per square metre. Heat outputs for different types and sizes of pressed steel radiators are given in table 5.1.

Table 5.1 Heat output from central heating radiators

Radiator type	$W/m^2{}^\circ C$	W/m^2 with water at 55°C above room temperature
Plane single panel	22-25	1200-1400
Plane double panel	35-40	1900-2200
Plane triple panel	48-54	2600-3000
Designs with added fins	27 - >150 depending on type	1500->8000

There are a number of choices to make when selecting pressed steel radiators. First among these is the number of panels. The simplest radiators are single panel. These are an appropriate choice for small radiators where a low heat output is required, and for larger radiators where the heat is to be disbursed over a wide area for even heating. A further advantage of these radiators is that their narrow profile allows them to be mounted quite close to the wall, which is occasionally a consideration.

Double panel radiators are also very common. These give approximately double the heat output per metre square of a single panel radiator. The extra heat output from a double panel radiator compared to a single panel is almost entirely convective rather than radiant. This is frequently claimed as an advantage for single panel radiators because radiant heat is felt immediately by the room occupants whereas convective heat, which relies on circulation of air currents, is slower to reach occupants. I find this argument a bit weak because even a single panel radiator has the vast majority of its heat output as convective. I have never noticed a difference in the comfort of adequately heated rooms with single or multiple panel radiators!

Triple panel radiators are available but are much less commonly used because in most rooms double panel radiators give ample heat output for the space they take up.

Single, double, or triple panel radiators are available in two types: convector or plane. Convector radiators have additional fins welded onto them that increased the heat output. The additional heat output provided by the fins is entirely convective and not radiant. The extent to which the heat output is increased depends on the design but broadly speaking they approximately double it. This means that a single convector radiator will have a very similar heat output for the same size as a double plane radiator. They usually also have very similar cost and take up the same amount of space! This means there may be little to choose between single convective and double non convective radiators but double non convective ones have an advantage that is surprisingly valuable in use: you can hang twice as many clothes over them to dry! This is because the space between the panels is not occupied by the convector system so towels and the like can hang down between the panels.

When deciding on radiators it is usually fairly clear whether a simple single panel, double panel convector, or something between is needed. If something in between is needed I would generally recommend a plane double panel for the rather prosaic reason mentioned above.

The final choice to be made is between "seam top" and "roll top". These terms refer to the way in which the steel sheets which form the radiator are joined together along the top edge. Figure 5.1 illustrates the difference. The seam top type are generally slightly cheaper and have a slightly higher heat output than the role top type but these

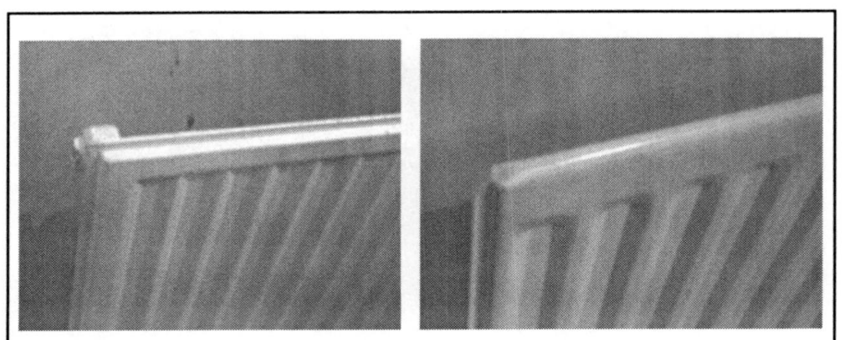

Figure 5.1: The two types of pressed steel radiator panel: a, seam top; b, roll top.

differences are pretty negligible. The roll top type offer a round more easily cleaned surface that is less likely to leave rust stains on fabrics left on the radiator to dry.

Cast-iron radiators

Cast-iron radiators were the standard installation with convective central heating systems. This was because iron casting technology lends itself to radiators with large water channels allowing adequate water flow under the conditions of very low differential pressure found in convective systems. Figures 5.2 and 5.3 show radiators of this type dating from the first half of the twentieth century in a Northumberland church. Pressed steel radiators have narrow channels with higher resistance and are less suitable for convective circuits. They are much cheaper to produce and have superseded cast-iron radiators since pumped circuits became standard. There are circumstances where convective central heating systems are in use and cast-iron radiators are commonly found. This is particularly true of public buildings. Cast-iron radiators are still manufactured and are available for replacement of existing radiators though this is a rare application because they are extraordinary durable and reliable.

Figure 5.2: A cast iron radiator dating from the first half of the twentieth century. You can just see that both ends are connected at the bottom

Figure 5.3: An early radiator formed of large horizontal pipes. The wide water channels offer low resistance to flow and are suitable for convective systems.

As well as offering a lower resistance to water flow, cast-iron radiators have a "period" appearance that is becoming increasingly popular. They work perfectly well with modern central heating systems and aesthetics is the usual reason for their being chosen over pressed steel alternatives. They are considerably more expensive for the same heat output and are also much heavier. They are usually floor standing with a bolt holding them on to the wall. They can be attached the wall without legs but this requires quite substantial brackets.

The period effect of cast iron radiators can be ruined by installing a modern thermostatic control valve. Period look valves are available but have to be specifically sought (figure 5.4). They are well worth using to complete the effect.

Radiator Valves

There are three types of radiator valves in common use: thermostatic, manual, and lock shield. Radiators are fitted with two valves, one at each end. Until recently the standard fitment was one lock shield and one manual valve. The manual valve was fitted on the flow side of the radiator and used to control its heat output. The lock shield valve was fitted on the other end and used to balance the radiator system. Manual valves offer rather limited control over the heat output of the radiator and it is now required in the UK that new installations are equipped

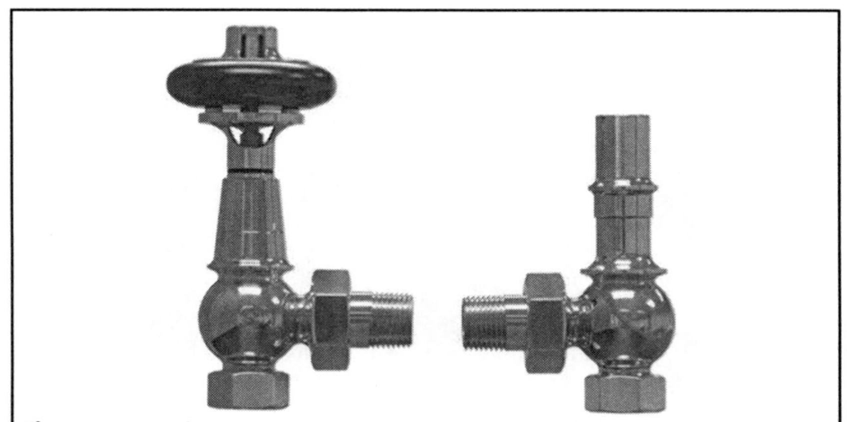

Figure 5.4: Thermostatic radiator valve and lock shield set styled to blend with the period look of cast-iron radiators. Reproduced with permission of Brass Products Ltd., Ashford, UK.

with thermostatic radiator valves. These can be fitted at either end of the radiator.

The flow and return connections to radiators can be fitter either at the top or bottom or one to the top and the other to the bottom. The control valve may be fitted at the top so that it is located at a convenient height. Gravity systems are different. Gravity fed radiators work best if the flow end of the radiator is connected at the top and the return end is connected at the bottom.

Lock shield valves

Lock shield valves are intended to be set by the central heating installer or service engineer and not tampered with by the users of the building. They are in all respects like simple manually operated valves except that there is no handle to adjust them and the spindle from the tap is covered by a plastic shield. This is removed to gain access to the spindle which is turned with a spanner to adjust the valve.

Thermostatic radiator valves

Thermostatic radiator valves (TRV) comprise two components: a valve and a thermostatic head. The valve is a tap opened by a spring. The tap has a pin sticking out of it. The valve closes when this pin is pushed in. The control part of the valve is a metal cylinder containing a mixture of liquid, wax, and gas that expands as it gets warmer. The cylinder comprises a bellows system which accommodates the expansion of the contents by getting longer, pushing the pin in. The control part sticks out into the room and takes the same temperature as the surroundings. As the surroundings warm up the liquid/gas/wax mix expands depressing the pin and closing the valve. A negative feedback system is established that controls the room temperature.

There are a couple of things to note about these valves. Firstly they respond rather slowly to changing temperature taking 20 minutes or so to react. Secondly because of the negative feedback system they spend most of the time in a marginally open position. This can lead to noise or knocking.

It is now a requirement that TRVs be fitted to radiators in rooms where there is no wall thermostat. In some installations all radiators have been fitted with TRVs. This means that in warm weather when the building is fully heated all valve are off. This can be undesirable as the pump can be on when all valves are closed. Differential pressure between the flow and return sides can become quite large and this may lead to knocking. One solution is to fit a differential pressure release valve. It is important to check the boiler type as condensing boilers should not be used with such valves. An alternative to a differential pressure release valve is to have one or more of the radiators without a TRV, using a manual valve that cannot be completely closed or a lock shield valve instead. Connected this way the radiator is always on. This radiator should be located in the same space as the thermostat, commonly the hall.

Twin entry radiator valves

Twin entry valves are radiator valves where the flow and return are connected at the same corner of the radiator. A long thin pipe runs along the bottom of the radiator to carry water to the far end giving flow through the whole radiator. They were used with micro-bore systems in the 1980s but are now obsolete.

Balancing Radiators

Modern central heating systems use radiators connected in parallel. The rate at which water flows from the flow to return sides of the circuit varies from radiator to radiator. The fastest flow occurs through the part of circuit that has the lowest overall resistance. This is generally the radiator nearest boiler because of the reduced pipe run. More distant radiators heat more slowly and reach lower average temperatures. This inequality of heating is countered by balancing the central heating system. Balancing takes time and care. Systems using thermostatic radiator valves work reasonably well in most situations without any balancing and it is regrettable that many if not most installations nowadays are not balanced.

Radiators are designed for a 10°C temperature drop from the flow to the return side. The temperature of flow from the boiler should not exceed 85°C and this falls to 80 or so by the time it gets to the radiators making the optimal return temperature of around 70°C. This gives a mean water temperature of 75°C. Condensing boilers use a cooler return temperature because returning water is used to condense steam from the boiler exhaust. When installing central heating systems it is usual to provide for an ambient temperature of 20°C in Bathrooms and 21°C in all other rooms. There has previously been a tendency to arrange for lower temperatures in Bedrooms and hallways of around 18°C but modern control systems allow much more accurate and specific regional control of temperature so this is no longer necessary. The slightly higher temperature of 23°C should be aimed at in buildings inhabited by elderly or infirm persons.

The objective of balancing is to adjust the lock shield valves so that all radiators have the same temperature difference between the flow and return sides when the system is fully on. There are two methods of balancing. The first involves trial and error. Turn all radiator valves to fully on and turn on the central heating system. When it has reached normal operating temperature after 20 minutes or so feel all the radiators. If one is notably the hottest close the lock shield valve by ¼ of a turn. After a further 20 minutes asses the radiator temperatures. If the same radiator still hottest turn it down a further ¼ of a turn. Repeat this process until another radiator is hottest, in which case start turning down its lock shield valve in the same way, and so on until all are the same temperature.

The second method is considerably quicker but the necessary equipment is quite expensive. It involves two thermometers to clamp onto the pipes on either end of the radiator and measure their surface temperatures. An advantage of this technique is that heat output is more accurately reflected in the temperature difference between the flow and return pipes of the radiator then it is by the surface temperature. The central heating system is turned on and all radiator valves opened. The temperature difference between the ends of each radiator is measured. The radiator with the lowest temperature difference is adjusted first. Aiming for a temperature difference of 10°C the lock shield valve is closed by a ¼ of a turn. It takes a few minutes for the temperature difference to stabilise again after the adjustment has been made but when it has a further adjustment is made if the difference is still below 10°C. When the difference stabilise at 10°C move on to the radiator that now has the lowest temperature difference and repeat the process.

The above procedures assume that the radiators are well matched to the rooms. If they are not it is possible to compensate for this by deliberately out-balancing them. If a radiator is too large for the room it can be made to run cooler by closing the balancing valve to have a temperature difference of more than 10°C with an average temperature of lower than 75°C. In the case of a room with inadequate radiators the reverse can be done, running the radiator with a smaller drop and higher average temperature.

Venting radiators

Mains water contains dissolved air. Unlike the solubility of solids, the solubility of gases in water is reduced at raised temperatures. This means that when a central heating system is filled with cold water a certain amount of dissolved air is introduced. When the water is heated this precipitates and collects at the top of radiators. This gas must be vented to allow the radiators to heat up to the top. This process of filling the system, heating it, and venting is part of installation and servicing. If air, or more strictly gas, subsequently collects at the top of radiators something is wrong. The possible causes are:

Hydrogen

Galvanic corrosion occurs in all central heating systems that have no inhibitor added or inadequate inhibitor levels. Hydrogen gas produced by galvanic corrosion is a cause for radiators needing to be frequently bled. The inhibitor status of the water in the central heating system can be checked either by sending a sample to a central heating service specialist or by using the two glass test described in chapter 6. The solution is to flush the system and add an inhibitor.

Micro leaks

Small leaks can let air into the system if they are located up stream of the pump where the pressure can become sub-atmospheric when the pump is running, especially on upper floors where the hydrostatic pressure is lower. These leaks may or may not allow water to leak out when the pump is not running. When they do it is a great help because the leak can be readily found by a tell tale rust stain or green deposit on a copper pipe; when not it may not be possible to locate the leak at all.

If this problem is suspected look closely at the pump connections first, particularly on the inflow side. If that is sound go round the radiators starting with the one nearest the boiler on the return side and working backwards. If a leak is found because of a rust stain tighten or remake the joint. If not tighten all compression fitting nuts by 1/8 turn, more if they feel loose. If that does not solve the problem it will be necessary to drain down the system and remake the screw thread and PTFE joints directly connecting with the radiator. Even this may not solve the problem because micro air leaks can occur in apparently sound soldered joints. When a soldered joint cools the solder does not all solidify at once. Centres of solid solder from and spread into molten solder. When the solder solidifies it contracts so that gaps can form where two spreading areas of solidification meet. These may be plugged with flux at the time of making and not appear until the flux has dissolved away maybe years later.

If a leak is found coming from one of the radiators itself rather than the joints to it, serious corrosion throughout the system should be suspected. The leaking radiator will have to be replaced and it is quite likely that all the radiators are badly corroded internally. As a temporary measure the radiator can be isolated by closing both the control and lock shield valves. The radiator may empty via the leak which will then stop but the rest of the system can still be used while arrangements are made to replace it. After the system has been flushed, refilled and an inhibitor added, numerous Micro leaks in radiators that had previously been plugged by a lime scale or corrosion products may become apparent. If a radiator has to be replaced because of a micro air leak the entire system should be carefully inspected when it is back in operation. Also users should be warned to look out for tell-tale signs of leakage such as the need for frequent bleeding of radiators or the appearance of rust stains.

In cases where micro leaks are suspected but cannot be located a leak sealing compound can be used. This is a product that is added to the circulating water in the central heating system and solidifies in leaks, plugging them. Its action is similar to the automotive product Radweld®. Many boiler manufactures advise against the use of such sealants and they should only be used as a last resort because they may narrow or block water channels in the heat exchanger.

Air in-drawing and pump-over

These faults can arise with open vented systems and are discussed further in chapter 2. Pump-over introduces gas into the system not because of air bubbles being drawn down the cold feed but because gases are readily dissolve in the cold water in the cistern. The constant circulation of water through this expose surface in contact with the atmosphere brings dissolved gases into the system which are then driven out of solution when the water is heated.

Under-floor Heating

This system is gaining popularity in the UK and has been used in domestic environments in other countries for many years. The principal advantages are aesthetics and comfort. There are no visible heating elements and building occupants find it pleasant to walk on a warm floor. The heat distribution is preferable to that from radiators as there is even rising of warm air across the room rather than warm currents above the radiators to the ceiling with corresponding cold currents across the floor toward the radiators.

Under-floor heating also has disadvantages. It is difficult and expensive to install in an existing building. The heating elements are reliable but if they do leek of fail it is complex and disruptive to repair them as they are buried under the concrete screed of the floor. There is a small cost in efficiency when the system is used on the ground floor as some heat is lost into the ground. A significant problem if the system is to be installed in an existing building is the height that must be added to the floor which effectively reduces the ceiling height of the room if the floor base cannot be lowered accordingly. This is approximately 150 mm or 6 inches.

Under-floor heating comes in two varieties: electric and hot water. The electric system is inexpensive to buy, fairly easy to install, and reliable but energy costs are high. It is most commonly used for small areas of flooring where under-floor heating is particularly desirable such as in bathrooms. The usual practice in such installation is to lay an electric heating element mat in the cement that holds tiles down.

The hot water under-floor heating system is more complex and expensive to install but has competitive energy costs. It is usually installed on the ground floor of a house. The structure of the floor is illustrated in figure 5.5. A layer of insulating foam block with a thermo-reflective foil coating on the upper side is laid on top of the structural floor. This is to minimise loss of heat into the ground. This insulating layer has to be quite thick (100 mm) to minimise losses. The reflective foil side of the foam blocks is placed uppermost and heating pipes are located on top of it. They are held in place with a series of plastic fixings before the concrete screed of the floor is applied. The heating pipes are quite sophisticated. They can be bent round fairly sharp angles without flattening and are covered with a plastic material which protects them from corrosion and allows for their thermal expansion. The layout of an installation of heating pipes prior to applying the floor screed is shown in figure 5.6. Ideally there should be no joints between lengths of pipe underneath the screed.

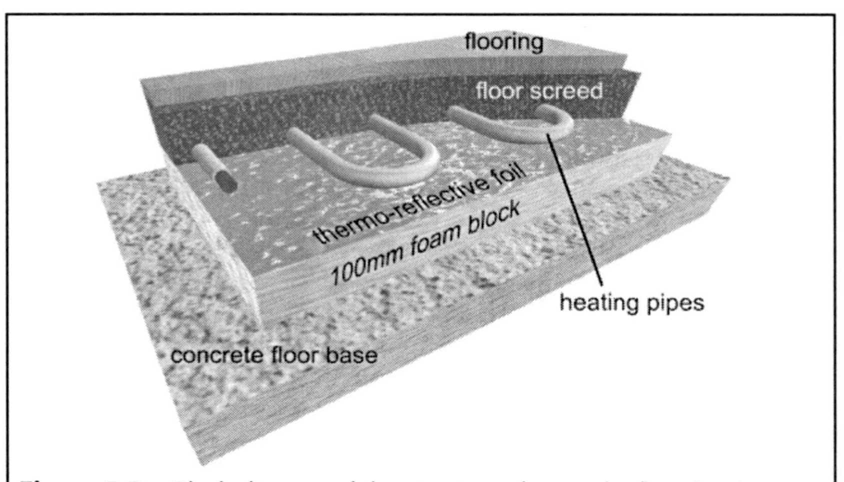

Figure 5.5: Block diagram of the structure of an under-floor heating system. The heating pipes are laid on insulating foam block coated with a metal thermo-reflective layer. The pipes are plastic coated to allow for expansion. The structure is about 150mm thick and so adds significantly to the height of the floor.

Figure 5.6: Wet under-floor heating system before the screed is applied. The heating pipes are placed above the thermo-reflective metal covered foam insulation that minimised heat loss into the ground.

Once the pipes are in place floor screeding is applied consisting of a specially formulated concrete mixture. It has to be applied with great care so that the pipes are not damaged. If a pipe should be injured by a clumsy shovel blow this may not become apparent until the heating system is pressurised after the screeding has set. It will then be necessary to tear up the screening over the leak and splice in a new piece of pipe with joints.

The choice of flooring is affected by under-floor heating because insulating material such as carpet reduces the effectiveness of the heating system and increases losses into the ground. Moreover wood flooring can be problematic because under-floor heating will dry it out

Figure 5.7: The central control point for the under-floor system part of which is shown in figure 5.6. There is a double bank of valves. The upper valves are manually operated and adjust zone flow rates. The lower valves are thermostatically controlled and respond to the temperature of return water.

to a lower water content than standard seasoned timber. This tends to make it shrink after installation revealing gaps.

Under-floor heating systems are arranged in separately controlled zones. At a central location a series of control valve is installed from which each of the zones can be regulated (figure 5.7). Wall mounted thermostats are used but there is no equivalent to thermostatic radiator valves within the room.

CHAPTER 6

Corrosion, Lime Scale and Chemical Additives

Corrosion

Corrosion can affect any part of central heating systems but its effect is usually most obvious in radiators. Corrosion occurs fastest when a suitable reactant, usually oxygen, is continuously added to the water. Sealed systems are protected from this but will still corrode so must also have a corrosion inhibitor added. The principal causes of corrosion are oxidation and galvanic electrolysis.

Oxidation is rusting in the case of iron or tarnishing in the case of copper or brass. It can only occur in the presence of oxygen dissolved in water. Open vented systems can be affected because oxygen dissolves in the water in the cistern. Sealed systems when functioning properly are almost completely protected from oxidative corrosion. The oxygen that is dissolved in the water when the system is filled is limited and if the system is not leaking no further oxygen will gain access to the water.

Galvanic corrosion is a feature of systems that involve different metals connected together by both an electrical conductor and an aqueous solution[1]. Such systems form an electrochemical cell (figure 6.1) where the solution is called the electrolyte and the metal objects dipped in it are called the electrodes. The reactions given in the middle column of table 6.1 occur on the surface of the electrodes. They convert metal atoms in the electrode to charged metal "ions" in the electrolyte or vice versa. In so doing they produce a voltage difference between the electrode and electrolyte that is different for

[1] An aqueous solution is a solution of chemicals in water. In this context it is a solution containing ions where ions are atoms that have an electric charge by either having electrons (e- in table 6.1) removed or added to them. This is relevant to corrosion because metal ions are vastly more soluble in water than uncharged metal atoms, so to corrode metal its atoms must first be "ionised". The ionic solution must be electrically neutral so for every positive metal "cation" there is an equivalent charge of negative "anions", common ones being chloride (Cl^-) and Sulphate (SO_4^{2-}).

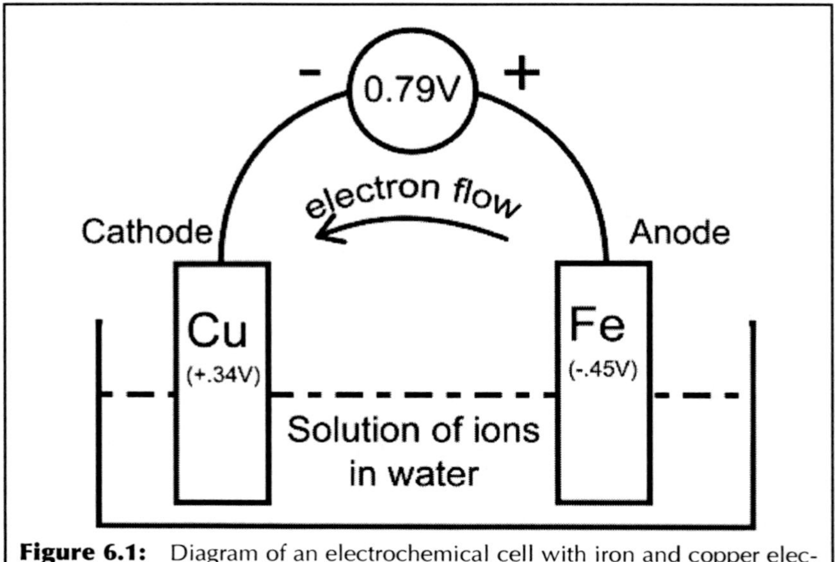

Figure 6.1: Diagram of an electrochemical cell with iron and copper electrodes.

different metals (table 6.1). Whether the reaction goes from atom to ion or ion to atom depends on the externally applied voltage between the electrode and electrolyte. If the electrodes in a cell are both made of the same metal there will be a difference between the voltage of the electrolyte and electrodes but it will be the same for each so they will be at the same voltage and nothing will happen if they are connected together. If they are not of the same metal they will be at different voltages. When connected together the reaction at one electrode will be forced into reverse and current will flow. This is the basis of electric batteries.

All electrochemical cells must have two electrodes so the voltage difference between a single electrode and electrolyte cannot be measured in isolation but the voltages between electrodes can be compared. Experimental measurements on cells using different electrodes have allowed a comparative series of metals and their voltages compared to a standard electrode to be drawn up. By convention the standard is the "hydrogen electrode". Table 7.1 gives the electrochemical series for the metals commonly found in central heating systems. Some metals appear more than once in the table because they have more than one type of ion, e.g. copper which has cupric (Cu^{2+}) and cuprous (Cu^{+}) ions.

As written in the table, ions in the electrolyte are combined with electrons from the electrode and deposited as metal. The electrode where this occurs is called the cathode of the circuit and the reaction "cathodic". On the other side electrons are taken from atoms in the electrode, ionising them and transferring them to the electrolyte. This electrode is called the anode and the reaction "anodic".

Table 6.1 The electrochemical series of metals used in central heating systems

Metal	Reaction	Voltage relative to Hydrogen
Copper	$Cu^{+} + e^{-} \rightarrow Cu$	+0.521
Copper	$Cu^{2+} + 2e^{-} \rightarrow Cu$	+0.3419
Hydrogen (acid)	$2H^{+} + 2e^{-} \rightarrow H_2$	0
Iron	$Fe^{3+} + 3e^{-} \rightarrow Fe$	-0.037
Lead	$Pb^{2+} + 2e^{-} \rightarrow Pb$	-0.1262
Tin	$Sn^{2+} + 2e^{-} \rightarrow Sn$	-0.1375
Nickel	$Ni^{2+} + 2e^{-} \rightarrow Ni$	-0.257
Cobalt	$Co^{2+} + 2e^{-} \rightarrow Co$	-0.28
Cadmium	$Cd^{2+} + 2e^{-} \rightarrow Cd$	-0.403
Iron	$Fe^{2+} + 2e^{-} \rightarrow Fe$	-0.447
Zinc	$Zn^{2+} + 2e^{-} \rightarrow Zn$	-0.7618
Chromium	$Cr^{2+} + 2e^{-} \rightarrow Cr$	-0.913
Manganese	$Mn^{2+} + 2e^{-} \rightarrow Mn$	-1.185
Aluminium	$Al^{3+} + 3e^{-} \rightarrow Al$	-1.662
Magnesium	$Mg^{2+} + 2e^{-} \rightarrow Mg$	-2.372
Magnesium	$Mg^{+} + e^{-} \rightarrow Mg$	-2.7

Galvanic corrosion occurs at the anode where the reactions of table 6.1 are reversed. We can work out which metal will form the anode by looking at table 6.1 and figure 6.1. Of the two metals the one lower down the series with the more negative electrochemical potential will form the anode.

In figure 6.1 the electrodes are iron and copper. They are dipped in an electrolyte. With a voltmeter connected between the electrodes, the voltage shown is the difference between the electrochemical voltages of the two metals from table 6.1. When the electrodes are connected together electrons flow from negative to positive[2]. At the iron (Fe) side electrons are leaving the block and being taken from Fe in the block to make Fe^{++} in the solution – corrosion.

At the copper (Cu) side the situation in practice is more complex because there are numerous ions in solution, particularly Fe^{++} and possibly a little Cu^{++}. The impression from the idealised explanation above that the copper block will get larger is an over simplification but it will not lose metal. One of the reactions that can be reversed at the cathode is the hydrogen reaction giving off H_2 gas. Another cathode product is a crust of mixed chemicals that have been "un-ionised" and forced out of solution and stuck to the cathode, or a sludge if not.

Of the metals commonly used in central heating systems the lowest in table 6.1 is aluminium above which are zinc and iron. Aluminium is used a lot in boilers. It is a very reactive metal but is protected from corrosion because very shortly after an aluminium component is manufactured a tough oxide layer forms on the surface which is impervious to most attacking chemicals and being a very poor conductor of electricity prevents electrolytic corrosion. Electrolytic corrosion of iron can be a problem but in general iron components are large compared to those of other metals higher up the series, particularly copper, so the rate of corrosion is relatively slow and small amounts of metal loss are well tolerated. A particular problem with iron corrosion is at the junctions between iron and copper piping. The adapter unions are usually made of brass (an alloy of copper and zinc). Galvanic deposition on the brass side of the union can narrow the pipe causing slow flow or even blockage. The deposits are easily removed with a file once the problem is diagnosed.

Galvanic zinc corrosion can cause a problem because although pure zinc is not used in most central heating system it forms a component brass. Galvanic corrosion leads to "dezincification" of brass. Zinc atoms are removed from the brass leaving a brittle, porous copper residue. This frequently affects the components of taps that are immersed in water leading to crumbling and failure.

Galvanic corrosion is a problem that affects a wide variety of fields including shipbuilding and civil engineering. It can be prevented by the use of a "sacrificial anode". The idea is that a large lump of a cheap metal which is low-down the electrochemical

[2] By a rather awkward convention electrons are negative and flow from negative to positive. This arose because electrical positive and negative were defined before the nature of electrons was known.

series is connected to the metal structure that is being protected and immersed in the same water. This lump slowly disappears but can be easily replaced. However this means that the other metals higher up the series are protected while the lump still exists. Magnesium is the metal commonly chosen for this purpose. Magnesium sacrificial electrodes are available for central heating systems but they are not often used. Instead chemical corrosion inhibitors are added to the circulating water.

Other possible causes of corrosion are certain chemicals used in the assembly of central heating systems such as solder flux, jointing compound, casting sand and swarf from pipe cutting.

Chloride

Excess chloride (Cl^-) in the circulating water promotes corrosive pitting of several metals including aluminium and stainless steel. These two are used in modern condensing boilers. Chloride is contained in many contaminants particularly fluxes and detergents and can be increased by poorly maintained water softeners.

pH

Aluminium is protected from corrosion by a hard oxide layer that forms on the surface of any exposed metal. This layer is stable in the pH range 4 to 9 but outside this range the oxide layer becomes unstable and rapid corrosion results. Most modern condensing boilers use aluminium heat exchangers so keeping pH within these limits has become important and this function is served by inhibitors.

Consequences of corrosion

In principle very severe corrosion can lead to leaks but these are comparatively rare and when they occur they usually indicate that corrosion is so far advanced that the effected component, or quite probably the entire system, is at the end of its life. More of a problem is not the loss of metal but the accumulation of the products of corrosion as a sludge which can block water channels and foul valves. Rust may accumulate at the bottom of radiators to the extent that the lower longitudinal channel may become blocked leading to cold patches.

Sludge normally settles downstream of where it originates. Small particles that remain in suspension can circulate around the whole system and contribute to scale formation. A particularly tenacious form of scale can form in the heat exchanged as a result of mixed lime and sludge deposition. Scale and sludge lead to failure of thermostatic radiator valves, pumps, motorised valves, cocks and heat exchangers.

Testing the central heating water for corrosive effect

If corrosion is suspected, a simple test can be done to see if there is a problem with the water circulating in the central heating and if so if this is caused by the chemical composition of the mains water supply. It is done with two clear jam jars that can be sealed from the atmosphere. Place a copper coin and iron nail in each one. Fill one jam jar with cold water from a tap in the house and seal it. Label it as mains. Fill the other jar with water bled from a radiator, seal and label it. Examine the jars after 1 week to see if there has been any corrosion of nails or copper. If there has and it affects both jars then corrosion is present and caused by chemicals in the mains water. If it is only present in the central heating jar then the system needs flushing and an inhibitor adding.

If the water drawn from the central heating system is found to be red it is an indication of oxidative corrosion due to oxygen entering the system. This has to be corrected before flushing and inhibiting.

Lime scale

Rocks of the limestone group are soluble in water and water supplies in areas with a large amount of limestone in the environment contain dissolved calcium bicarbonate. With most solid materials (as opposed to gases) their solubility in water increases as the temperature rises but in the case of calcium bicarbonate, increasing the temperature increases conversion to insoluble calcium carbonate which then precipitates as lime scale.

Water with a high lime content is known as hard water (table 6.2). Lime scale deposition increases with harder water, the temperature of the water, and the alkalinity of the water. Lime scale principally affects the hot water system because unlike the central heating system there is a constant supply of new lime bearing water. Is a serious matter causing early failure of a wide range of components, particularly, boiler heat exchangers as this is where the water is hottest. In areas where water hardness exceeds 200 milligrams per litre it is generally recommended to fit a base exchange water softener in the incoming mains supply. This reduces lime scale build up in all plumbing and kettles in the building. Base exchange water softeners reduce mains pressure by about 1 bar so a mains pressure of 4 bar or more is necessary to provide adequate cold water pressure. Where a water softener is fitted it is a legal requirement in the UK to provide un-softened water for drinking to at least one tap.

Table 6.2. Classification of water hardness

Designation	mg/l of Lime Scale (CaCO3)
Soft	0-50
Moderate Soft	50-100
Mild Hard	100-150
Moderate Hard	150-200
Hard	200-250
Very Hard	>250

A system affected by lime scale can be restored with the use of a commercial chemical lime scale remover or descaler. Descalers are made up mainly of acids such as acetic or citric acid. They are limited in the amount of lime scale they can dissolve and care is necessary to maximise their effect. They usually come in a solid form that is dissolved in water. To descale a hot water system it is first emptied by tying up the float valve in the attic cistern and opening all the hot taps. When they stop running the taps are closed and the system filled with the descaler solution. It is only filled as far as the inlet to the cold feed running from the cistern to the bottom of the hot water tank so that there is only a shallow layer of water in the bottom of the cistern. This is to limit the amount of lime scale remover solutions that is required (there is no scale in the cistern as the water there is always cold – or should be). The whole system is then left for the scale to dissolve over a period of time. The descaling chemical within pipes is more quickly exhausted than it is within the tank so small amounts of solution are run out of the taps at regular intervals to refresh the chemical. Exact details are given by the manufacturers of the descaling agents but the whole process takes several hours and thorough flushing afterwards is necessary.

A controversial device is the electronic lime scale remover. It emits radio waves from cables wrapped around or protruding into the rising main pipe of the house that are supposed to alter the form of calcium containing micro-crystals in the water. This is said to keep calcium carbonate in suspension and limit its deposition on surfaces. The problem is that no clear mechanism of action or convincing evidence of benefit has reached the peer reviewed scientific literature. Among many other hyped water treatment products, they appear to be an application of pseudoscience.

Central heating and softened water

Base exchange water softeners tend make the water more alkaline which destabilised the oxide layer that protects aluminium from corrosion. Water softened by water softeners is used in the secondary hot water circuit but should never be used for the central heating circuit of systems containing aluminium, which increasingly they do. If the system does not contain aluminium an inhibitor specifically designed for softened water may be used but in practice it is safest never to use softened water.

Biological contamination

Biological organisms such as algae, bacteria, fungi and yeasts can grow in central heating systems. The relatively cool cisterns are prone to this if not properly installed. The main risk is from legionella (see chapter 4) but other problems such as slime blocking narrow water ways and a form of anaerobic bacteria that promote corrosion of ferrous metals are occasionally encountered. If microbial contamination is suspected the system should be treated by chlorination. Modern inhibitors control all of the above problems. All systems except those with primatic hot water cylinders must be inhibited but prior to that they must be cleaned.

System cleaning

Cleaning is an essential part of central heating system commissioning, maintenance and repair. Previously fresh water flushing was often used alone for this purpose but since 2006 the British Building Regulations (part L) have required that chemical cleaning agents are used. Two cleaning methods are in use: gravity and power flushing. As the name suggests, gravity cleaning involves filling the system with the cleaning agent and then draining it down by gravity. Power flushing involves a high-pressure pump that forces cleaning fluid at speed round the circuit. Cleaning is a slow process that cannot be rushed. It takes time for the chemical agents to have their effect and care to ensure that all areas of the system are adequately exposed to the cleaning fluids. Power flushing typically takes 4-6 hours for most domestic systems and gravity cleaning 8 hours. The cleaning processes is continued until water in the system is running out clear. This can be estimated by visual inspection as many contaminants cause visible colouration or clouding. Better is to use an electronic instrument known as a total dissolved solids meter or TDS meter. This enables the amount of contamination in the central heating water to be estimated as a percentage of that in the inflowing mains water. Some of the explanations below assume that a TDS meter is available.

Power flushing is quicker and more effective than gravity cleaning but requires expensive equipment, though it is readily available for hire. In general the dirtier the system and the more lime scale and sludge expected, the greater the argument of power flushing. In badly corroded systems it may be that corrosion debris is blocking pinholes in radiators and flushing could lead to leakage by removing the debris. In this situation the radiators need replacing, ideally before power flushing but the problem frequently does not become apparent until afterwards. Certain types of system are not suitable for power flushing. These include single pipe systems and systems with stainless or passivated steel piping[3]. Special care is needed with some systems or components and reference to manufacturer's data is recommended. These include gravity hot water systems, primatic cylinders, thermal stores, twin entry radiator valves and micro-bore pipe work.

Some preparatory work is necessary before power flushing. To begin with ensure that the system is free from leaks. Any remedial work requiring the system to be drained down should be completed before cleaning. If the system is open vented the cold feed and vent should either be joined together temporarily or capped of for the washing process. All electric controls should be switched off. All radiator lock shield valves should be open and thermostatic radiator valve heads removed. Diverter or zone valves should be set to their manual open positions and one-way valves such as antigravity valves should be bypassed or temporarily removed as the flushing system uses flow reversal to optimise cleaning.

Power flushing machines are pumps that are connected to the central heating system in two places. This allows the machine to pump water around the system. A facility exist to reverse the flow of water and most power flushing machines contain a reservoir to which chemicals can be added. Some systems can be used with the boiler on for a hot flush as chemical cleaning is faster at higher temperatures.

The basic sequence is:

1) Start with all radiator valves open. The flush is run for 15 minutes or so with water only, reversing the flow direction every couple of minutes. The water is then drained out and replaced with clean until the Total Dissolved Solids meter reading of the effluent is within 20% of that of the mains water supply.

[3] Passivation is a treatment used on stainless steel to reduce corrosion. It involves exposing the surface to an acid that removed some iron atoms and promotes the formation of a protective oxide layer.

2) Add the selected cleaning chemicals to the power flushing machine's reservoir.
3) Flush the primary hot water circuit for 10 minutes, reversing the flow direction several times.
4) Return to the space heating circuit. Close all radiator valves except the most distant from the point where the power flushing machine is connected. Flush this radiator for at least 5 minutes with several reversals of flow direction. If a hot flush is being used keep flushing until the radiator is equally hot at all points. During the flush strike the radiator with a mallet to mobilise adherent debris within it. Depending on the level of contamination it may be necessary to drain off and replace the flushing fluid. Do this until the TDS meter reading is within 10% of that of the mains.
5) Turn off the valves to the now flushed radiator and open those of the next nearest to the flushing machine. Repeat step 4 with this radiator.
6) Repeat steps 4 and 5 with each radiator in turn working towards the machine.
7) Turn off the boiler if a hot flush has been used and drain out the cleaning water replacing it with mains until the effluent TDS reading is within 10% of the mains water.
8) Repeat the whole process from 1 to 7 doing the radiators in the reverse order, i.e. moving away from the machine and ending with the primary hot water circuit and a final flush of the space heating circuit with all radiator on.

Gravity cleaning and flushing

Before starting, if the system is open vented tie up the float valve in the expansion cistern. Then drain down the system completely. It is helpful to measure the volume of water that comes out as this eases the process of flushing. Make the necessary repairs while the system is drained down. Ensure that full bore drains are available at low points in the system and that they are accessible. If covered they should be under easily removable panels and their location marked. If drains are not full bore, inaccessible or not at the bottom of the system fit appropriate drain cocks.

The flushing sequence is:

1) Refill the system with mains water and add the selected cleaning chemicals. The quantity of chemical is often a function of the volume of water contained in the system hence the value of measuring the volume of what comes out when the system is drained down. There are several locations at which the chemicals can be added. In an

open vented system they can be added to the attic expansion cistern. Alternatively they may be added directly into one of the radiators or at any other convenient connection. Manufacturers provide specific plumbing devices to do this.

2) Turn on all radiators and the primary hot water circuit.
3) Now turn the system on. The cleaning agent manufacturers will generally specify the temperature at which the operation should be carried out and how long for. It may be as long as a week.
4) In the case of an open vented system switch off the cold water feed or tie up the float valve in the cistern.
5) Drain down the entire system from a low full bore drain cock.
6) When water has stopped coming out of the drain cock open all radiator vents starting from the top of the house and moving downwards. Drain the entire system to dryness.
7) The system must now be rinsed with fresh cold water. This involves closing all the radiator vents and refilling the system. Each vent is then reopened in turn to allow air trapped in the radiators to escape. The system is then run to circulate water round for a few minutes.
8) The complete drainage process from 4 to 7 is then repeated. It takes at least three full flushes to rinse the system. If a TDS meter is available rinsing should continue until the effluent shows a TDS within 10% of that of the incoming mains water.
9) Add the inhibitor either to the cistern if there is one or into a convenient radiator with a suitable injection device.

Inhibitors

Corrosion inhibitors are chemical products that are added to the circulating water of central heating systems. Commercial products combine many different inhibitors to inhibit a number of different corrosion reactions in different metals. They reduce the rate of corrosion by interfering with both the cathodic and anodic corrosion reactions. The commonest type of inhibitor is "primarily anodic" or "passivating". These react with the metal to form a corrosion resistant coating that is maintained while the inhibitor is present in the water but is quickly lost if the inhibitor is used up or flushed out. When an anodic inhibitor is used the dose should be enough to form the coating and leave enough inhibitor in the water to maintain the coating. If too little is used it will form and incomplete coating and leave none in solution, in which case corrosion protection will be not just reduced but vastly reduced leading to an on off effect. Protection form corrosion is only afforded if the amount added to the system is over a certain level as specified by the inhibitor manufacturer. Below that level it is largely ineffective, adding too much makes little difference and does not improve protection.

As well as slowing corrosion inhibitors also reduce sludge formation as this is a product of corrosion reactions, and inhibit lime scale formation.

Several chemical that may contaminate central heating water can adversely affect the performance of inhibitors. These include solder flux, detergents such as cleaning agents, washing up liquid, washing powder and existing corrosion sludge. It is good practice, and now required by the British Building Regulations, that systems be cleaned and rinsed by flushing before the inhibitor is added after installation or work requiring draining down. A total dissolved solids meter or commercial tests kit can be used to test for cleanliness before adding an inhibitor. Inhibitor should be added immediately after cleaning the system as without an inhibitor lime scale will quickly form in the boiler heart exchanger reducing efficiency.

It is difficult accurately to estimate the amount of inhibitor needed because of the variability of system layouts so it is sensible to check that the level is adequate after it has been added and run round the system. Inhibitor manufacturers make test kits to do this or offer a laboratory service for the purpose. Such tests should be repeated annually at the time of boiler servicing. Inhibitors come with a sticker to attach to the boiler on which to write the date the inhibitor was added or last checked and the next check due date. Bear in mind that overdosing will not adversely affect the system but under-dosing will greatly reduce protection.

Chapter 7
Plumbing

Working with central heating systems requires knowledge of some aspects of plumbing. Fortunately the types of pipe commonly encountered are easy to deal with and the skills required accessible to the competent DIY installer. In many parts of the world, including the UK, regulations require that gas installation is done by qualified persons registered with a government agency[1]. It is a safety precaution because of the life-threatening risks that can arise if gas appliances are improperly installed or maintained. This is a relatively minor imposition because only a small part of the plumbing involved in central heating system is directly connected with the gas supply. A registered engineer is only required to install the boiler. The entire system beyond that including all hot water to taps and radiators, with the exception of unvented hot water cylinders, can be installed by any competent person. The regulations regarding the installation of oil fired boilers are slightly less stringent[2].

Modern domestic central heating systems are almost exclusively plumbed with copper piping. Iron piping is encountered in older systems and in some specialised applications for which reason iron pipe plumbing is described below.

Copper pipe

A range of diameters of copper pipe are in common use in domestic heating systems: 28mm, 22mm, 15mm, and smaller diameters collectively known as micro-bore. In general direct connections to the boiler are made with 22mm copper pipe which is run to each floor in the house. The distribution of space heating and hot water

[1] In the UK the scheme is known as CORGI (Confederation for the Registration of Gas Installers) It was formed in 1970 in response to a disaster in 1968, when a 22-storey block of flats in London was destroyed by a gas explosion that claimed 5 lives.

[2] In the UK OFTEC (Oil Firing Technical Association) maintains a register of persons competent at installing oil fired appliances. The work must be done by a competent person but they do not have to be OFTEC registered though if not the local authority must be informed so that they can carry out an inspection.

around each floor is usually done with 15mm pipe which is cheap, easy to install, durable and has adequate flow capacity for most purposes.

The prevailing recommendation is that no more than three radiator should be run of a single section of 15mm pipe and it is reasonable to comply with this recommendation when installing a new system but an additional radiator can normally be added to a 15mm run without any problems connected with pipe diameter arising.

Microbore piping is becoming increasingly common. It is generally 12, 10 or 8mm in diameter. It is easier to install than larger diameter piping but its smaller cross-section area means that flow through it is more restricted and only one tap or radiator is supplied from each run. The system uses manifolds that are connected to the 22mm main pipes to the zones or floors and feed into several microbore runs. These manifolds are quite expensive! Microbore piping is very easy to manipulate around corners and requires no specific tools for bending except for sharp angles. It is sold in rolls rather than lengths which are easy to transport and store.

Bare Copper rapidly tarnishes and can be unsightly for which reason chromium plate is often preferred for pipes that are visible such as the parts of the flow and returned pipes to radiators that are above the floor. It handles very similarly to copper pipe except it is much more difficult to join with solder as the chromium plating has to be ground off first. This is not an easy job to do by hand. A mechanical abrasive such as a belt sander, angle grinder or grindstone is far quicker.

Bending copper pipe

Routing copper pipe round corners requires either the use of pre formed knuckles or some form of bending. Very gentle bends around a radius of 2m or more for 15mm pipe and 4m for 22mm pipe can be made with no tools at all just over the knee. More commonly a bending device will be needed.

The problem with bending pipe is its tendency to flatten. When this happens the copper walls on the inside the outside of the bend flatten and come together and the sides of the bend spread out and form a kink. This can be prevented and the pipe kept round around the bend by one of two means. Firstly something can be placed inside the pipe to prevent the inside and outside walls of the bend coming together. This is how bending springs work. Alternatively something can be placed outside the pipe to prevent the sides of the bend from spreading apart and this is how tube bending machines work.

Bending springs

Bending springs are cylindrical springs which have a similar outside diameter to the internal diameter of the pipe for which they are intended. They are usually about 500 mm long. They are very inexpensive, costing considerably less than a length of the pipe for which there are designed. They come with a hook at one end and an eye at the other so they can be linked together and fed into a pipe to allow a bend to be made a considerable distance from the end. Alternatively they can be pulled through the pipe with a cord. The technique of their use is to feed them into the pipe and position them so that the whole of the section of pipe that is to be bent is filled with the spring. The bend is made over a knee on a similar soft surface. If numerous bends are to be made a knee it can become quite painful! The system is inexpensive and works reasonably well but is rather inaccurate and bends cannot be made around as small a radius as is possible with a tube bender. It is difficult to make bends within about 300 mm of the end of the pipe because a length is needed on either side of the bend to provide leverage.

Tube benders employ a system of pulley like wheels with high walls that constrain the sides of the pipe as it is bend to prevent them spreading outwards (figure 7.1). They are more expensive than bending springs and range from costly high quality floor standing units to less expensive hand held devices. They are easier to use and allow curves of accurate radius and angle to be bent and this makes accurate pipe forming easier. They can bend tighter curves than a spring and place bends nearer to the end of a pipe because there is no need to use the pipe itself as a lever.

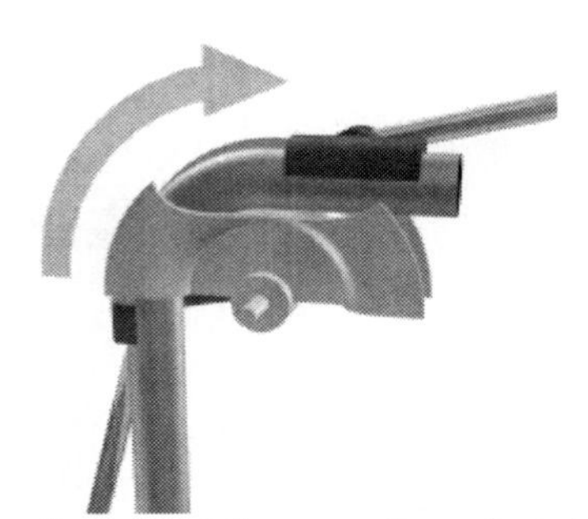

Figure 7.1: Cut-away diagram of the operation of a tube bending machine. The tube to be bent is placed in a pulley with high side flanges. A lever bends the tube over the pulley hub. The high flanges prevent the tube spreading and kinking.

Cutting copper pipe

Copper pipe is most easily cut with a pipe cutter (figure 7.2). This is a small device with 2 rollers and a cutting wheel. The pipe is located on the rollers and the cutting wheel driven into the opposite side of the pipe with a screw. The cutter is then rotated around the pipe so that the cutting wheel leaves a score. The wheel is tightened further into the pipe and the cutter rotated around it again. This deepens the score. The process is repeated until the pipe is cut through. Cutters of varying quality are available. The cheaper ones have the problem that they tend to spiral along the pipe as they are rotated rather than run true making a single circular score. This problem is not great with 15 mm pipe but is much more significant with the wider 22 and 28 mm sizes. If you are going to be cutting a lot of 22 or 28 mm pipe I would recommend purchasing a high-quality cutter.

A disadvantage of this cutting system is there has to be enough clearance around the pipe to allow the swing of the cutter. Special small cutters are available to allow pipe installed in restricted placed to be cut but they still require clearance of 40 mm or so. Figure 7.2 is an example of this type. Even this amount of clearance is frequently not available in which case alternative cutting methods are necessary. A simple hacksaw will sometimes suffice but has its own clearance problems especially as pipes are often installed in groups of more than 1 making difficult cut only one of them with a hacksaw. A small narrow cutting disc on a small grinder can be useful for this purpose in tight corners.

Figure 7.2: A small pipe cutter. This one is designed to allow it to be rotated round a pipe in a restricted space. A swing of 40mm or so is still needed. It is easy to imagine why this type is referred to as a knuckle breaker.

De-burring

Cutting leaves some copper burr on the ends of the cut pipe. Pipe cutters leave a burr on the inside which narrows the entry to the pipe. It is good practice to remove this burr with a de-burning tool (figure 7.3) as it prevents small fragments of copper coming loose in the process of corrosion and fouling valves or other control systems. A bending spring cannot be slid into the pipe passed the burr left by a pipe cutting wheel making deburring essential if one is to be used.

Joining copper pipe

The most durable, most reliable, and cheapest way of joining copper pipe is to solder it together with specially made fittings. It is no surprise therefore that almost all copper pipe joining is done this way. With modern materials it is very easy to do. There are two types of fitting available. These are solder ring and capillary (figure 7.4). Capillary fittings are a lot cheaper and give just as good a result and I would recommend their use. The disadvantage is that you have to buy a role of solder as well as the fittings but the overall cost is less than for solder ring fittings.

Several types of solder are available. Traditional plumbers solder is an alloy of 50% tin and 50% lead. This is still one of the cheapest types of solder and works very well but it is no longer permitted for plumbing pipe work that conveys drinking water. Alternatives to this are low lead and lead free solder. Low lead solder has a high melting point and is rather more difficult to use than 50/50 solder. Lead free solder is considerably more expensive but is easier to use than either of the other types. If plumbing is being done with the water supply system as well as the central heating system I suggests using lead free solder for everything as it avoids the possibility of using the wrong kind of solder and ease of use increases productivity which more than pays for the additional cost.

All copper surfaces to be joined must be clean and untarnished. Surfaces should be cleaned with a fine abrasive such as fine emery paper or coarse to medium steel wool. This is easy to do on the ends of pipes as the abrasive is simply run around them. With 22 mm fittings the sockets are wide enough for the abrasive to be run around the inside of the fitting on a finger. With 15mm or smaller fittings this cannot be done and a special wire brush is needed (figure 7.5). These brushes are available for 22mm pipe also but are unnecessary. They are rather expensive for what they are and they don't last all that long but there is really no alternative to their use with the smaller pipe sizes.

Figure 7.3: The business end of a copper pipe deburring tool. The tool is about the size of a pencil. The cutter is swept round the inside of the cut end of the pipe to separate the burr left by a pipe cutter.

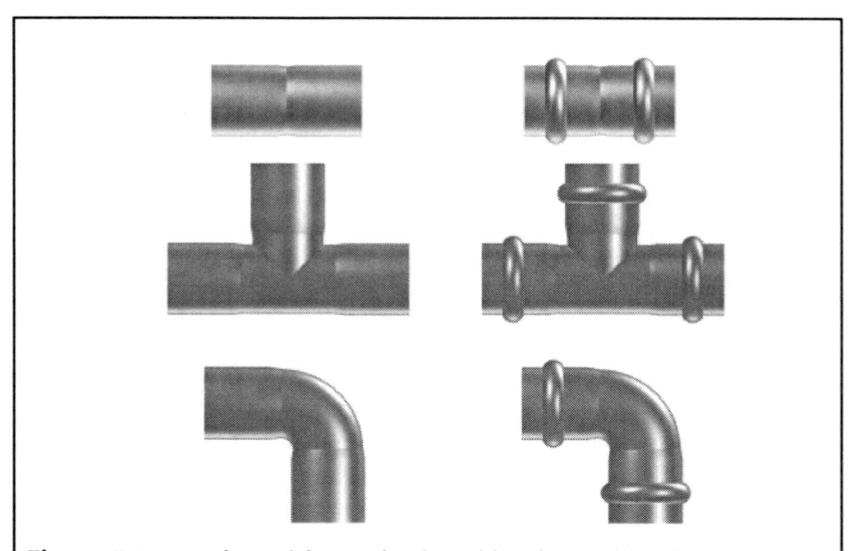

Figure 7.4: Pre formed fittings for the soldered assembly of copper pipes. Capillary fittings (left) are plane whereas solder ring fittings (right) have a solder containing bulge in the copper around each joint.

Figure 7.5: Wire brush designed to clean the inside of copper pipe fittings before soldering. This one is for 15mm fittings. It is reaching the end of its life as can be seen from the rounding of the top.

Once the surfaces have been cleaned with abrasive, flux needs to be applied to all joint surfaces. Flux is a chemical agent that cleans and deoxidises the copper surface. It is corrosive an irritant to skin so should be handled carefully. The best solution for flux application I have found is to use a small bristle artists brush with the handle shortened so that it fits inside a cylindrical cigar tin, protecting it while still wet with flux between soldering joints.

With flux applied the joint is assembled ready for soldering. If you are using solder ring fittings you simply assembled a joint and heat it with a blowlamp. After a few seconds the solder melts and is drawn into the joint by capillary action. You see a silver line of hot metal appear around the joint which tells you that the correct temperature had been reached. Melting points vary from one type of solder to another but are usually around 200°C. If using capillary fittings the process is very similar except that you have to add the solder while heating joint. Train the blowlamp on the joint for a few seconds and dab the corner between the pipe and fitting with the solder until it starts to melt. When the end of the solder wire starts to stick to the copper keep heating and blobs of solder will melt on the surface of the pipe and be drawn into the joint. Add solder to the joint until the same silver line of metal is seen all around. Be sure that the joint does not move while the solder is cooling.

It may take several seconds for the solder to cool enough to solidify especially if you have been enthusiastic with the blowlamp. The solder in the joint does not all solidify at the same moment and if the joint moves while the solder is partly solidified it is likely that a dry joint will result which will leak.

To allow the blow lamp flame to scorch whatever is on the far side of the joint is simply poor workmanship and risks starting a fire as joints are often made next to easily ignited dry wooden laths or timbers. Avoid this by arming yourself with a series of flameproof protectors such as piece of ceramic tile or metal and flame proof fabric squares that are available for the purpose. Sopping wet newspaper can be packed against whatever needs protecting and is useful in a tight corner. A wet cloth is handy not only for flame protection but also for wiping excess solder or flux and for cooling a joint quickly after it has set so it can be touched. Be prepared to put out a fire if you do start one.

Problems with soldered joints

Soldered joints are generally very easy to make and very reliable. It is quite possible for a DIY plumber to assemble an entire central heating system with hundreds of joints and find no leaks when it is first turned on. There are problems that can arise and correcting them after the event is much more difficult than getting it right in the first instance for reasons that are explained below.

Dry joints are alluded to above. A dry joint arises when the solder satisfactorily coated both surfaces to be joined but cracks have arisen within the solder as a result of the joint being moved when only partly solidified. The usual situation is that the solder around the surface of one or other pipe solidified first so that any movement leads to a cylindrical crack formation. The way to avoid this problem is to be careful not to move the joint before the solder has set. If for some reason particularly vigorous heating has been necessary it is prudent to wait a generous amount of time - perhaps ten seconds or more - before moving the pipe work. If you feel that a joint has moved in the cooling phase the solution is to reapply the blowtorch until the solder melts and then let it cool again this time without movement.

When one of the surfaces to be joint is not properly cleaned the solder will not stick to the copper and a leak is likely. This problem is overcome by adequate cleaning. It usually arises in situations where an existing pipe is being soldered to a new connector and the side of the pipe nearest the wall cannot be seen. The solution to this problem is to wrap an abrasive around the pipe and clean it with a bowstring action. That way if the part of the pipe you can see is clean the part you can't almost certainly will be.

Sometimes the joint gap is too large for capillary action to act properly. This happens when the joint is assembled dry and then struck a heavy blow with a hammer for example so that both the fitting and the copper pipe are distorted together. It may be necessary to do this deliberately in some circumstances where pipe work has to fit into a very restricted space. If only one of the bits being assembled is off round enough to cause a soldering problem the joint will not fit together so the problem doesn't arise. If you aware of the problem before hand it is often possible to solder by simply being rather generous with the solder. An alternative is to put pieces of copper braid or wire into the widened gap so that the overall space is narrower allowing capillary action to draw solder right across the defect. This is a trick for repairing holes in pipes that typically arise by nails or drills being inadvertently driven into them. If it is difficult to expose enough pipe to splice in a replacement peace the problem can be solved by cleaning the surroundings of the hole and packing in a small piece of copper wire or braid then heating the hole up and soldering.

Wet pipes

Wet pipes are the real curse of soldered joints. They can present such a problem that it is just not with the hassle of dealing with soldered joints and despite the expense and poorer durability, compression fittings are chosen instead. The problem with wet pipes is that the specific latent heat of vaporisation of water is so enormously high. If you have a copper pipe with water inside it, it doesn't matter how vigorously and long you heat it with a blowtorch it will never get hotter than 100°C until all the water has boiled away. If water keeps running into the pipe from somewhere else you will never be able to reach the melting point of solder. Assembling systems from soldered copper pipe is pretty easy especially if the pipe is new. Making changes in an existing system that is already full of water is a very different game for this one overriding reason. The first priority when making soldered joint in an existing system is to ensure that the thing you're joining is free from water and this is likely to be considerably more difficult than either cutting the pipes to length or preparing and assembling the joint. There are a number of technical devices which may help.

Freezing the pipe

Freezing bursts pipes because when water freezes the ice takes up more volume than did the water. Pipes burst when a length of water is trapped between two frozen section and the water then also freezes. With the outlets blocked by ice the volume increase stretches the pipe and may split it. It will not happen if only one point in the pipe is being frozen. Freezing a pipe blocks it temporarily. That allows work to be done without draining the whole system. There are of course problems. Central heating systems contain chemical that lower the freezing point. Clamp on electric freezing gear is expensive. Spray on freezing is a cheaper alternative but does not automatically keep removing heat. The freeze has to be a couple of feet from the joint or the heat from the blow lamp will melt the ice. This may lead to the freeze failing or to water dripping down and preventing the joint getting hot enough to solder.

Draining the system

The traditional approach to plumbing work on central heating systems is to drain them down first. This is a safe, easy and relatively low cost option, the cost being in replacing the inhibitor. It is not quite such a sure solution to the problem of wet pipes. The complexities of central heating circuits means that when a system is drained down it will still not be fully empty. Water will trickle down through the pipes at a dribble for a considerable time. The manipulations necessary to assemble a soldered joint are liable to provoke such a dribble on the inside of the pipe that will make soldering impossible.

There is a problem with the behaviour of flux that aggravates the problem of wet pipes. When you heat a joint the flux liquefies, flows over the surface and attacks any oxidation leaving bare copper ready for solder. If the solder is not applied immediately however the flux tends to evaporate and when this happens the oxidation rapidly reforms because of the high-temperature. You therefore do not have an option of heating a joint for a long time before applying the solder. This means that if there is a little water in the pipe you can't simply boil it away with the blowlamp and then solder the joint. It is not necessary for the pipe to the bone dry but anything more than a little surface moisture will preclude satisfactory soldered joints.

Compression fittings

Compression fittings are considerably more expensive than solder fittings and they are not as reliable with a tendency to leak if expose to physical knocks or after long time periods. The equipment and materials necessary for soldering are not required which is an advantage but the main advantage in central heating systems is that compression fittings do not require the pipes to be dry. The time and effort needed to ensure that joints stay dry for soldering is so large that compression fittings are a more economic option for connecting to an existing system. Suppose for example you're adding an extension circuit of radiators to an attic floor. A very reasonable approached to this would be to assemble the entire circuit with new pipe and soldered joints but to use compression fittings for the T junctions that connect onto the existing central heating system.

Iron pipes

Iron piping is rarely if ever selected for domestic heating installations nowadays but may occasionally be encountered in the older gravity driven systems. Working with iron pipes is a more complicated business than working with copper. They cannot easily be bent so angles must be provided with elbow fittings. They are connected together with threaded joints. The process requires fairly heavy gauge equipment and specifically made fittings. The best way of cutting iron pipe is with a ferrous pipe cutter. This is a large-scale version of the type of copper pipe cutter shown in figure 7.2. Ferrous pipe cutters leave a slight raised lip on the outside of the cut end. This is a great aid in commencing the later thread cutting process. A hacksaw can be used but does not leave this useful lip.

A male thread is cut on the end of the pipe with a thread cutting die. Thread cutting oil has to be added approximately every other revolution of the die. When cut, the thread is covered with PTFE tape to about 4 layers. A pipe jointing compound is spread on the female side of the threaded joint which is then screwed together and tightened.

CHAPTER 8
Heat

Basics of heat and energy

Heat is not an easy concept to define so it is fortunate that we have an intuitive idea of what it is. It is that property that an object has more of when it is hot than when it is cold. It is also fortunate that heat behaves in a simple way that is easy to understand and predict.

Heat is a form of energy (another difficult concept to define). Quantities of heat are measured with the same units as quantities of energy. The SI unit (Système International d'unités) of energy is the Joule (J). In terms of domestic heating, 1J is not very much. It takes about 100,000J to heat enough tap water to boiling point to make an ordinary mug of coffee. For this reason the kJ (equal to 1000J), MJ (equal to 1000,000J) and GJ(1000,000,000J) are more commonly used.

Most heating calculations involve the rates at which Joules are given off by a heater or absorbed by its surroundings rather than total numbers of Joules. Such rates are expressed as Joules per unit of time. The SI unit of time is the second (s) and rates of heating are expressed in J/s. The J/s is such a widely used unit that it has its own name: the Watt (W). Rate of energy conversion is called power so the unit of power is the Watt. As with the Joule, the Watt is rather small so we usually quote kW (1000W).

A somewhat perverse unit is the kilowatt hour (kWh). Like the Joule this is a unit of energy but it is expressed in terms of power multiplied by time. The unit is used because it is easy to reckon total energy usage of appliances by multiplying their power in kW by the time for which they are in use in hours. 1kWh = 1000W×1 hour =1000W×3600 seconds or 3600,000J (or 3.6MJ).

Temperature is another awkward quantity to define. Object A is hotter than object B if heat flows from A to B. The rate at which this heat flows is proportional to the difference between the "hotness" of A and that of B. The term "temperature" refers to this degree of hotness and is measured in degrees. Fahrenheit (°F), Centigrade (°C) or Kelvin (K – notice that the

° symbol is not used for Kelvin) are in common use. Converting between °C and K is fairly simple as the units are the same size but the scales start at different temperatures. Absolute zero (0 on the Kelvin scale) is -273.15 °C so temperature in K is temperature in °C+273.15. Fahrenheit is more complicated because the units have a different size and the scale starts at a different temperature. Zero °C = 32.00°F and one degree on the centigrade scale is equal to 1.800 or 9/5 degrees on the Fahrenheit scale. Degrees C = (°F-32)×5/9 and °F = (°C×9/5)+32.

Before the SI units were introduced the English speaking world used the BTU (British Thermal Unit). The heating industry has been particularly slow to change and the old BTU is still frequently encountered in the UK and is in general use in the USA. The BTU is a unit of energy like the Joule. It is the energy required to heat 1 pound of water by 1°F. Originally this is how it was defined but now it is defined as 1055.05585262J. Rather confusingly it usually appears not as the BTU of energy but as a unit of power. It would be nice if this unit of power was the BTU per second as it would then be close to 1kW but that, of course, would be too easy. It is BTUs per hour and the unit is usually abbreviated to BTU. What is more the BTU per hour is not much power (=0.293W). Domestic heaters are in the range 5,000 – 150,000 BTU per hour so quite often 1000 BTU per hour is also abbreviated to 1 BTU! It is usually obvious from the context what BTU is meant.

Specific Heat Capacity

The specific heat capacity of a material is the amount of heat energy required to raise the temperature of a unit of mass of that material by one unit of temperature. In SI units this is the energy in J required to raise 1kg by 1K[1]. The word "specific" in this context means per unit of mass or weight[2] (the kg in S.I. units). The heat capacity of an object is equal to the specific heat capacity of the material from which it is made multiplied by its mass. The volumetric heat capacity is the heat capacity

[1] Strict scientific definitions use degrees Kelvin (K) rather than degrees centigrade (°C). As the size of the units is the same it makes no difference which scale is used when working with differences between temperatures and it is generally convenient to stick to familiar °C for practical purposes.

[2] Mass and weight are not the same thing. Mass is a measure of the amount of matter in an object and is defined by the rate of acceleration the object experiences when subject to a particular force. Weight on the other hand is the force of gravitational attraction on the object, which is proportional to its mass. Weight therefore varies with local gravity whereas mass does not. On Earth 1kg of Sugar has a mass of 1kg and a weight of 1kg. On the moon 1kg of sugar has a mass of 1kg but a weight of 165g! Strictly all properties referred to here are per unit of mass not weight but on earth they are virtually the same.

per unit of volume rather than mass and is equal to the specific heat capacity times the density of the material. It is more useful than the specific heat capacity when quantities are reckoned in volumes rather than weight as is the case for most masonry. Both specific and volumetric heat capacities vary slightly with temperature but not enough materially to affect our calculations as a high degree of accuracy is not required. Some useful heat capacities are listed in table 8.1.

There are various things to note from table 8.1. The first is the very high heat capacity of water, both specific and volumetric. Another is that all types of masonry used in building have similar heat capacities of about 840 J/kgK or 1950 J/lK. For calculations it is usually good enough to assume these values for all masonry. The exception is soapstone, a rock

Table 8.1 heat capacities and densities of some materials used in buildings

Material	Specific Heat capacity J/kgK	Density kg/l	Volumetric heat capacity j/lK
Dry air at 20°C	1160	0.00120	1.4
Water at 20°C	4180	1	4180
Bricks and mortar	840	2.2 - 2.4	1850-2020
Building stone	840	2.5	2100
Concrete	800	2.4	1920
Glass	820	2.6	2130
Granite	800	2.7	2160
Asbestos	840	1.6	1340
Marble	880	2.6	2290
Soap Stone	980	2.5-2.8	2450-2740
Cast Iron	545	6.8-7.8	3700-4250
Steel	517	7.5-8	3880-4140
Copper	395	8.9	3520
Brass	395	8.4-8.7	3320-3440
Lead	129	11.3	1460
Aluminium and alloys	900	2.5-2.8	2250-2520

with an unusually high heat capacity and also a high heat conductivity. It is very soft. Its two main applications are in teaching sculptors to carve and in making fireplaces and stoves.

Latent Heat.

Heat energy must be transferred to change the state[3] of a substance. It takes 420kJ to raise 1kg of water from 0 to 100°C but to turn it into steam at 100°C requires a further 2.26MJ! This amount of heat is called the specific latent heat of vaporisation of water. It is of relevance to fuel combustion because virtually all fuels release water vapour when they burn. This water comes from hydrogen within the fuel which is oxidised to water or trapped water which is vaporised. If this water vapour is condensed to liquid water it gives up 2.26MJ/Kg in latent heat. It is this heat that is recovered by condensing boilers. The calorific value of a fuel can be calculated with this latent heat (gross) or without it (net). Common usage in the USA is of gross calorific values whereas in Europe net values are used. For this reason American figures for calorific values are higher than European and efficiencies of appliances are correspondingly lower.

Heat requirements of buildings.

Accurate assessment of the power required to heat a room or building adequately depends on a knowledge of the thermal conductivities or "U" values of the walls and ceilings and the expected difference in temperature between the inside and the outside. Such data will be available in the planning and construction stages of new buildings but is not likely to be to hand thereafter, particularly for old buildings. In these cases estimates are required.

As a first simple estimate most domestic rooms in temperate climates require between 0 and 6kW to heat them depending on the time of year. A better estimate is that in temperate climates average domestic rooms require up to 100W per m^3 of room space.

A more accurate estimate still can be obtained from the formula:
heat requirement in kW = (80A +300B + 20CD)E
where:

A = area of wall or ceiling in contact with the outside in m^2
B = area of glass in m^2
C = number of room air changes per hour
D = volume of room in m^3
E = temperature difference between the room and outside in °C.

[3] "State" means solid, liquid, or gas. Changing state means melting, freezing, boiling, condensing, or subliming (solid to gas). Latent heat must be added to melt or boil something and removed to freeze or condense it.

This formula contains terms for the loss of heat by conduction through the walls, ceiling and windows (80A and 300B) and another term (20CD) which refers to ventilation losses.

A comfortable environment requires ventilation of living spaces by the constant inflow of fresh air and outflow of stale room air. When heating rather than air conditioning are in use the fresh air is cooler than the desired room temperature and must be heated. The heating power required to warm this incoming fresh air is referred to as ventilatory loss and as with other powers is expressed in Watts. Calculating ventilatory loss is fairly straightforward. It is simply air flow in l/s × the temperature difference in °C × the volumetric heat capacity of air at atmospheric pressure: 1.4J/ °C. As in the formula above air flow is often expressed as the number of times the air in the room is replaced each hour or "air changes per hour". The requirement for comfort in most rooms most of the time is a ventilation rate of ½ to 2 air changes per hour.

As an example I will calculate the ventilatory heat loss from a typical domestic room. Suppose the room is 65m^3 (a moderate size) and is ventilated with 2 air changes per hour. The outside air temperature is -3°C and the room temperature is 20°C. What is the ventilatory loss?

The amount of air flow is 65×2 = 130m^3/hour or 36 litres/second.

The heat lost per second is therefore 36×(20-(-3))×1.4 J/s =1163W.

If the outside temperature falls to -20°C, not at all uncommon in northern latitudes in the winter, this figure rises to 2022W.

Open fires induce considerably more ventilatory air flow than is required for comfort and this is one major source of their inefficiency. Figures are typically between 2 and 10 air changes per hour.

If it is important to make an accurate estimate of heat requirements in cases where the above calculations are likely to be misleading, for example in odd shaped rooms or particularly exposed buildings, then an experimental approach is possible. This is based on the observation that the heating power required is proportional to the difference between the temperatures of the room and the outside air. The method is to heat the room in question exclusively with heaters whose heat output in Watts is known. It is usually most convenient to use electric heaters for this purpose as they can be assumed to be 100% efficient and their power consumption in W is printed on the appliance. Ensure that all other forms of heating are either of known power output or are turned off. Note the total power required to keep the room comfortable. Add on 140W for each adult in the room[4] while the experiment is in progress to get P,

[4] 140 W is the heat output form an adult not engaged in heavy exercise or sleep.

the total heating power. Measure the inside and outside temperatures. The outside temperature is T and the comfortable room temperature is R. Then consult meteorological statistics for the area to find the coldest outside temperature at which the room is likely to have to be adequately heated. This temperature is called C. The power required during such conditions can then be calculated as

Maximum heating power required = $P \times (R-C)/(R-T)$

Even this can be in error if for example high winds significantly affect the cooling of the outside of the house or induce cold drafts.

Supplying heat to a room

Once the heat requirement of the room has been estimated we can choose the heating appliances. When using direct electric heating it is a simple matter of adding the power consumption figures on the appliances to get the required amount but this form of heating is unlikely to be economic. The alternatives present greater difficulty in calculating the heat output.

In the case of central heating radiators, outputs can be calculated from the manufacturer's data and the temperature of the water which averages about 75°C for conventional and 65°C for condensing boilers. Table 6.1 gives a guide to the heat output per m^2 of radiator area for various types. The total output when filled with water at 75°C in a room at 20 °C is also given. Fifty five to 85°C is a typical range for the flow temperature of water from a boiler in a domestic central heating system.

Stoves have outputs specified by the manufacturer though these will assume that the appliance is operating at maximum efficiency which is generally only achieved when the output is near its maximum and the stove is in good repair. Stove efficiencies are between 40 and 75% for solid fuel fired, greater for gas or oil fired.

Open fires

The increase in air flow and ventilatory heat loss induced by open fires can be unpredictable. Worse is that the efficiency of fireplaces is so variable that the heat output can be equally difficult to estimate. The effect of ventilatory heat loss can bring the efficiency of an open fire below 0% and 40% is about the limit for the most efficient designs. This means that choosing a fireplace for a room is inevitably more hit and miss than choosing other heating appliances. If an open fireplace is to be installed provision should be made adequately to heat the room entirely by other means.

Fireplaces may be inefficient but that does not mean that they are not powerful. An open fire with an efficiency of 25% burning ordinary

Table 8.2. Approximate heat output and room sizes for coal fires		
Grate width	Max. power	Approximate room size m3
400	6kW	35-75
450	7.5kW	50-100
500	9.5kW	75-125
500	11.5kW	100 +

coal at a modest rate of 1kg per hour will provide 2kW of heat to the room and lose 6kW up the chimney. A grate 400mm wide by 225mm deep (a standard size in the UK) can comfortably consume 2.5kg/hour and dissipate 5kW into the room and 15 up the chimney. The problem is not so much that heat production is inadequate but that it is expensive in terms of fuel use and unpredictable in that one fireplace with and efficiency of 40% will dissipate 8kW into the room for the same amount of fuel burned as another with 10% efficiency which dissipates only 2kW into the room.

Table 8.2 gives a rough guide to the heat output into a room and approximate room size that can be adequately heated by the 4 commonest sizes of coal fire used in the UK. These are very rough figures. The lower limits placed on room sizes are arbitrary as there is no meaningful lower limit on how small a fire can be burned in any size of grate.

Wood burning fires are different again because the heat output of wood is only around a third that of coal for the same volume of fuel. A wood burning fire will need to have a grate about 1.4 times as wide (and so have about twice the area and contain 3 times the volume) for an equivalent heat output to a coal fire. The larger fireplace opening means greater ventilatory heat loss but this may be partly controlled by a fireplace throat damper (a hinged metal panel that can be moved to close the throat and reduce air flow up the chimney). The size of wood fire needed adequately to heat a room is less predictable than for coal but broadly a grate width around 1½ times those given for coal in table 11.3 will be about right.

Chapter 9
Motorised Valves

Valves operated by an electric motor are now standard components in central heating systems as they allow more sophisticated and accurate control than would otherwise be possible. Such "motorised valves" come in three different types. The simplest have two connections: an inflow and outflow. They are referred to as two port valves. More complex are three port valves (figure 9.1) that come in two types: diverter and mid position. A diverter valve has a common connection or port labled AB. This is connected to one or other but not both of two other ports, A and B, by the motor action. The third type of motorised valve is the mid position valve. This is a diverter valve with the additional facility to allow the common to be connected to both ports at the same time, its "mid position" function.

Figure 9.1: One of the most popular types of three port mid position valve – the Honeywell V4073. The ports are marked A, B and AB on the brass casting. Port AB is the common and is connected to the flow from the boiler. Port A is normally closed and open when the motor is energised. It is connected to the space heating circuit. Port B is normally open and is connected to the primary hot water circuit.

Motorised valves are sold as complete units but they are formed from two components which can be fairly easily separated. These are the plumbing bit (figure 9.2) and the electrical bit (figure 9.3). Under the plate of the plumbing part a rubber sphere is carried on an arm connected to the spindle. This is moved between two pipe openings to close one or the other, or remain in the middle with both open (figure 9.4). Most motorised valves use 22 or 28mm compression fittings for the plumbing side which has a maximum operating pressure of 8 to 10 bar. The operating temperature range is usually between 5 and 95°C.

The electrical bit requires a mains electricity supply. Power consumption is usually modest being under 10W. They are supplied with a length of heat resistant flex. Switching the valve takes a few seconds. It typically takes about 20 seconds to move the valve in the direction driven by the motor and 10 seconds for it to move back to its normal position driven by a return spring.

Since the mid-1990s motorised valves have been built so that the electrical actuator part can be replaced without having to disconnect any of the plumbing or drain down the system. This is a great advantage as the electric actuator is the part that most commonly fails. With many models the individual components of the actuator can be replaced separately. The actuator contains the motor and gearing with a return spring and a manual override lever which is used for filling and draining the system (figure 9.5).

Installation of motorised valves is fairly straightforward with a couple of things to note. The motorised actuator bit should always be above the plumbing bit so that any leaks do not result in water flowing into the electrical actuator and shorting it. Also there should be enough access available to make it possible to remove the actuator and replace it.

The wires to most motorised valves now conform to a common colour coding scheme. Details of the colour coding and connections are given on page 146.

Diverter valves

Diverter valves are used to switch the boiler output (port AB) between the space heating and hot water systems. They do not have the facility to provide heat to both systems are once. Almost all systems are hot water priority so that when both systems are calling the heat, the diverter valve is switched to the primary hot water circuit (port B) and will only switch to space heating (port A) when the hot water call for heat is satisfied.

The actuator system has an energised and de-energise state. If it fails it will generally fail in the de-energised state and this is arranged to be the one that is connected to the priority system. For this reason when

Figure 9.2: The plumbing part of the valve shown in figure 9.1 with the motor removed. The spindle turns an arm bearing a rubber sphere shown in figure 9.4. The bearing of this spindle may stick if the valve has not been operated for a long time. In that case it may be possible to free the spindle by turning it with a spanner.

Figure 9.3: The motor and switches of the V4073. The manual lever can be seen protruding from the casing under the flex.

Figure 9.4: The plate bearing the spindle has been removed and is seen from below. The arm bears a rubber sphere which is driven by the motor to block the opening to port B or by the return spring to block the opening to port A

Figure 9.5: The manual lever of the V4073. It is currently in the manual position. The lever moves slowly with sustained pressure against the return spring. It is then retained in the mid position behind the notch in the casing.

a motorised valve fails it is usually with the boiler output connected to the hot water circuit. A manual override lever is provided that puts the valve in a mid position with both ports connected to the common.

Diverter systems can perform poorly particularly on cold days when there is a high demand for hot water because no heat is given to the radiators until the hot water tank is completely heated. As this can take some time there can be inadequate space heating. The system has the advantage that as the boiler is only required to heat hot water or radiators and not both, a smaller boiler can be used. The system is not favoured now. More sophisticated systems with powerful boilers are preferred.

Mid position valves

Mid position valves allow the boiler heated water to be diverted to the space heating circuit, the hot water circuit, or split between the two. In order to allow a fairly even split of flow between the two systems they both have to be working at approximately the same pressure. For this reason mid position valves are only used with fully pumped systems (i.e. the pump is on when hot water is being heated) as party pumped systems have very different circuit pressures for the water and space heating. Like the diverter valve, they have an energised and a de-energised state and when they fail it is usually in the de-energised state. The de-energised state is usually used to connect to the hot water circuit. A manual override lever is provided that puts the valve in its mid position.

Two port valves

Two port valves are used to automatically switch some sections of the heating system on and off. When they are used the space heating is zoned into separate circuits. These zones may be the first floor, the ground floor and perhaps a further zone for a conservatory. The individual zones each have a two port valve on their flow side. When two port valves are in use the hot water circuit is also served by a two port valve and is treated as another zone rather than having a three port valve divert water between hot water and central heating and then have a series of two port valves breaking up the central heating system into zones. The valves are each controlled by a thermostat located in their zone or in the case of the hot water valve, a cylinder thermostat.

Two port valves are available as "normally closed" which are closed when de-energised or normally open which are open when de-energised. Normally closed valves are the standard type for most domestic heating systems. A manual override lever is provided to open them when de-energised.

Replacing the actuator

To replace the actuator the electrical control system should be turned off and the manual override lever left in the de-energised position. On modern motorised valves the actuator then simply unscrews and can be removed and replaced like-for-like. On older valves it may be necessary to drain down the system to the level of the valve because the plumbing components have to be opened in order to replace the actuator. If this is necessary an upgrade cover plate can often be purchased to allow easier replacement of the actuator in the future. Replacement is simplest if the exact same model of actuator can be purchased but if not it may be possible to fit a different actuator rather than replace the whole valve. The leading manufacturers now all use similar colour coding systems for the wires (chapter 14).

The plumbing part of motorised valves may fail. Leaks can occur at the junctions as is the case with any compression joints and the remedy is usually obvious. The control valve may seize. If this happens it should prompt an inquiry as to why. It may be that the valve has not operated for many months, perhaps over summer. A seized valve can often be freed by turning the spindle with a spanner once the actuator has been removed. If the valve has seized because of corrosion or debris within the system this must be corrected or other components will rapidly fail.

Chapter 10
Pumps

As pumps go, those fitted in central heating systems are fairly simple with few moving parts (figure 10.1). They are of the centrifugal type which have the advantages of being inexpensive to manufacture and having a long and reliable life. A disadvantage is that at low flow rates they are inefficient in energetic terms but in the context of heating systems the "lost" energy is negligible and in any event goes to further heat the water so is not wasted. The pressure raised by the pump depends on the flow rate which in turn depends on the resistance of the circuit. The pressure falls with increasing flow (figure 10.2) This means that as radiator valves close or parts of the circuit are switched off, the pressure in the remaining open circuit increases.

Four variations on the basic theme of centrifugal pumps are in general used. These are fixed speed, three speed, variable speed, and twin head. The simplest pumps are fixed speed. These have a pressure to flow relationship as shown in figure 10.2 line ③. They have the advantage of simplicity and were the standard for many years. When installed in systems with modest flow requirements they generate more pressure than is necessary leading to noise.

Most pumps sold now are three speed. They have three settings which correspond to the three different pressure / flow charts shown in figure 10.2. The central heating control system is not usually able to change the speed of the pump. This is set manually to be optimal for the system.

Selecting the speed is a matter of balancing between noise which occurs when speed is too high and inadequate flow which occurs when it is too low. The lowest speed at which the radiators are adequately heated is selected. To select the pump speed, first open fully all of the radiator control valves. Set the pump to its maximum speed and turn on the central heating. When the boiler starts the pump should also start. After about 15 minutes all the radiator should be fully hot. Turn the pump down to its minimum speed. After waiting 10 to 15 minutes if the radiators are all still fully hot then the minimum speed is adequate and should be left. If some of the radiators have dropped their temperature then set the pump

Figure 10.1: Exploded view of a Peglar Terrier® three speed central heating pump with the terminal block cover removed.

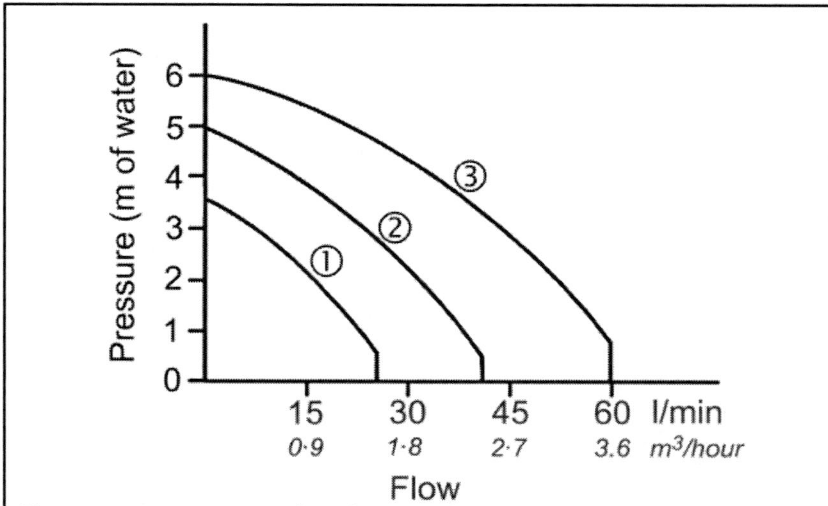

Figure 10.2: Pressure – flow diagram of a typical domestic three speed central heating pump. Flow is given in l/minute and m^3/hour. One l/minute = 60l/s =3.6 m^3/h.

to its second or middle speed. After waiting a further 10 to 15 minutes assess the radiator temperatures again. If they are all hot leave the pump on the middle speed, if not set the pump to its highest speed.

Pump speed can be used to keep an old system going a bit longer when it is in need of flushing. Fast moving water tends to keep sludge and air bubbles in suspension and prevents them from settling causing blockages. If this condition is suspected then start the pump at its maximum setting and only turn it down if excess noise is produced.

Noise

Pumps cause noise by several mechanisms. When in operation pumps vibrate slightly at the mains frequency of 50Hz. This is normally barely audible but is magnified if the pump is attached to, or touching, a panel that can act as a sounding board. Transmitted vibration to pipe-work is another common cause of noise. Several factors promote this. One is the pipes to and from the pump being under tension. Another is sharp bends in the areas of high flow rate particularly close to the pump.

The use of cushioning material between pipe work and joists or beams reduce noise as does a proper fixing bracket for supporting the circulating pump, many of which come with a facility to use sound absorbing rubber.

The bearings of a circulating pump when it is in good condition run virtually soundlessly but as it reaches the end of its life some play appears and this can lead to whining or grinding bearing noise. The solution is to replace the pump

Air entering the pump can result in noise that sounds like crackling or white noise. Air or gas within the system provokes failure of numerous components and should be investigated and eliminated. A dry running pump will be damaged quite quickly.

Cavitation is occasionally a problem in central heating circuits. It produces a white noise and rapidly reduces pump life. It occurs because the boiling point of water is significantly reduced at pressures below atmospheric. Within the pump there is an area of low pressure on the inlet side that is maximal along the rear surface of the impellor blades. If the pump is at a high setting and the hydrostatic pressure is close to or below atmospheric then pressure can drop substantially below atmospheric in this area. As the water circulating in the system is quite hot the pressure can drop low enough for it to boil. Steam bubbles then appear on the rear blade surfaces. The conditions of low pressure are highly localised and as the bubbles expand they reach higher pressure areas and quickly collapse. This formation and collapse of bubbles is quite violent, damaging the pump and causing noise. The solution to the problem is to lower the pump speed, or raise the static pressure in the system if this is possible.

Boiler kettling is not directly connected with the pump. It occurs when water passes through the boiler heat exchanger too slowly so the temperature reached is higher than it should be. Kettling makes a sound similar to that made by an electric kettle as it heats up a prior to reaching boiling point. The mechanism is described in chapter 11. There are various causes including obstruction of flow through the heat exchanger by lime scale or sludge or reduced flow can be a consequence of pump malfunction or too low a speed setting.

Automatic variable speed pumps

This kind of pump uses a feedback system to give constant pressure over a wide range of flow rates. They should not be used with systems were all radiators have thermostatic control valves as this can result in low flow rates through the boiler which can lead to boiler damage or kettling. A further type of system that these pump should not be used with is any combination boiler. Some combination boilers use variable speed pumps within and these should not be fitted with a bypass. The boiler manufacturer's instructions will make the requirements plain.

Automatic fixed speed and variable speed pumps

These pumps are the same as automatic variable speed pumps except that they also have the option of being operated as fixed speed usually with three selected speed settings. The difference with an ordinary three speed pump is that at each setting the actual speed of the pump is determined by feedback system that keeps the pressure constant rather than as shown in figure 10.2. In practice the two types are pretty much equivalent. Automatic pumps have a slightly more stringent ambient temperature requirement than other types.

Twin head pumps

Twin head pumps are a recent innovation. They are intended as replacements for a pump plus 3 port mid position valve. They have a common inlet to both pumps each of which has its own outlet. One is intended to pump the space heating system and the other the primary hot water. The control system turns one or other pump or both on to supply water to the two systems. There are no additional three way valves. The system has the advantage that the pump's characteristics can be matched to the resistance of the space heating and hot water circuits. The flow characteristics of the two are different and this has the advantage of reducing noise or improving heating response times.

Choice of pump size and setting

Most central heating systems are designed around a 10°C temperature difference between the flow and return sides of the space heating circuit. From the data given in chapter 8 it can be calculated that the flow rates to maintain this temperature must be 1.43l/minute per kW power output from the boiler. Pump flow rates are usually expressed in m^3 per hour and this equates to 0.086m^3 per hour per kW. For a 22kW boiler that makes 1.9m^3/hour (31.5 l/min) and 3m^3 per hour (50 l/min) for a 35 kW boiler. These are high flow rates. The cold tap of the kitchen sink in most houses when fully on will deliver 15 – 20 l/min.

Detailed calculations of pump flow rates are not normally necessary because pumps come in two sizes which are designed for heating systems with boilers of up to 22 kW for the smaller size and up to 35 kW for larger size and the three speed options give enough flexibility to match the pump output to the system in all but exceptional cases. Pumps come with charts giving the appropriate setting for boiler power and automatic flow control pumps come with a maximum boiler power specification. This usually means it's a simple matter to choose the right pump for the system.

Temperature control

Like all electric motors central heating pumps get hot while running. They are already kept hot by the hot water flowing through them and it is necessary to be careful not to insulate them. Rags or towels should not be draped over the housing or they will impair cooling. Pumps typically have an operating temperature range of 10 to 110 °C. They operate with static pressures of between about 0.5 and 10 bar over atmospheric and are factory tested to at least 15 bar. To prevent condensation within the motor windings the water temperature should be above the ambient air temperature while the pump is in operation.

Electric Supply and control

The electricity requirements of pumps are usually fairly modest being under 100W. A pump overrun control is sometimes used. This is a timing delay switch which keeps the pump running for a while after the border has switched off. This is an advantage with older cast iron boilers. Their water jackets have substantial heat capacity and overrun keeps water circulating through the boiler to prevent heat in the jacket walls from boiling the water.

Pump Position

The pump should be positioned on the flow side after the cold feed and vent connections but before any of the zone circuits or primary hot water circuit branch off the main flow pipe coming from the boiler. The effects of the pump on pressures round the circuit are complex. They are explained in chapter 2 under hydrostatics and hydrodynamics. For most purposes a detailed knowledge of these effects is not necessary but it is required to understand the malfunctions of the wide range of correctly and incorrectly installed systems that are encountered.

Pump location in a sealed system circuit

In a sealed system the location of the pump is not critical. Many boilers that are intended for use with sealed systems incorporate both the pump and expansion vessel within the boiler casing and this simplifies installation. If the pump is not included with the boiler then it is usually installed immediately following the boiler in the flow side though it works equally well on the return side after the last return connection (which should be primary hot water if an indirect cylinder is used) has joined.

Siting.

Circulating pumps use water to lubricate the bearings at either end of the spindle and so when installing care has to be taken to ensure the both bearings remain wet. For this reason they should be sited so the motor end of the spindle is either horizontal or somewhat higher than the pump end. It should not be lower. Neither should it be vertical or near vertical. An exception can be made to this last requirement in the case of a sealed system operating at a minimum of not less than 1.5 bar, but even this is not recommended.

If a pump is positioned on a vertical pipe it should pump upwards so that any air bubbles that pass through the pump will flow up with the water. If the pump is pumping downwards there will be a tendency for air bubbles to float up against the direction of water flow until they reach the high velocity flow area of the pump where they will be trapped. If it is essential to install a pump which is pumping downwards an air purger and air vent should be installed on its inlet side but this solution is not ideal as the air vent can itself allow air to enter the system. Better is simply to avoid pumping downwards.

Pump replacement

When replacing a pump, disconnect the electric supply by turning off the mains double pole switch to the heating system. Disconnect the pump flex. The system will have to be drained down unless isolator valves are present. If they are, turn them to isolate the part of the circuit containing the pump and drain only that. Before the pump is removed note the direction in which it is pumping and the way up it is fitted! First remove the vent plug (figure 10.3) and drain any water into a vessel. A jam jar is ideal for this because the colour of the water can be seen. If it is cloudy or coloured the system should be flushed and an inhibitor added. The two union nuts holding the pump to the system can then be undone and the pump removed.

Pumps are connected using a system of nuts and gaskets. When replacing pumps the gaskets should not be reused.

Venting

When a pump is newly installed or replaced it is essential that it is vented before being started. To do this the system should be turned off but pressurised. The vent plug is removed with a coin. Expect some water spillage. Beneath the vent plug is the end of the motor and impeller shaft. In it there is a slot to receive a screwdriver. Insert a screwdriver into the slot and wind it backwards and forwards to release any air trapped within the impeller.

Pump maintenance

Sometimes a pump that has not been used in many months, for example over the summer when the central heating was not in use, can fail to start because of accumulation of sludge. This problem can be solved by removing the vent plug and inserting a screwdriver blade into the slot in the shaft and turning it to free it. Running at the maximum speed for half an hour after doing this helps free up the Barings. The interior structure of a pump is quite simple as shown in figure 10.1 and they can be dismantled by removing four bolts holding the motor onto the casing. This will open the water way so the system should be drained before it is done. The components of the pump can then be inspected and cleaned.

Figure 10.3: View of the motor end of the pump seen in figure 10.1. The arrow indicates the vent plug.

CHAPTER 11
Boilers

It requires a considerable amount of expertise to work on boilers safely. This is reflected in many national regulation schemes. In general the more potentially dangerous the fuel used by boilers, the tighter the regulations controlling who may work on them. Because of this, in the UK work on gas appliances must be done by registered persons. Work on oil burning appliances must be done by competent persons. They need not be registered though this is encouraged. If they are not registered their work has to be inspected by a qualified and registered person. Less stringent requirements apply to solid fuel burning appliances. The purpose of this chapter is not to give the necessary expertise to do such works but to give enough background information on boiler types and operation to allow sensible choices to be made and appropriate instructions to be given to qualified installers.

In the early years of their development, machines that heated water were aimed at the production of steam for use in steam engines and industrial processes so they became known as boilers. The name stuck even though central heating boilers heat water to a temperature below the boiling point.

Selecting the boiler is one of the key decisions when installing a central heating system. The size of boiler required, or more accurately the power of boiler required, will be dictated by the heating needs of the building. Similarly in most cases the fuel used will be determined by local availability and cost. The principal decision left will be the type of boiler. The most commonly used options are combination, standard and system boilers. Boiler installations that are also cookers or "ranges" are occasionally chosen.

Condensing and boiler efficiency

The efficiency of a boiler refers to how much of the heat released by combustion of the fuel is given to the water. Greater efficiency is desirable for several reasons. It means less fuel is used to heat the same amount of water and therefore less expense. It also may minimise environmental damage by limiting carbon dioxide production. This is particularly important when fossil fuels are being burned, as they usually are. This latter reason has led to increasingly tight legislation about the acceptable performance of boilers.

Gas and oil are hydrocarbons. The molecules of the fuel contain carbon and hydrogen. The products of complete combustion of the fuel are carbon dioxide and water which is given off as water vapour. Water has an extremely high specific latent heat of vaporisation (2.26 MJ/kg). This latent heat can be recovered from the exhaust of the boiler fire and used if the water vapour is condensed back to liquid water. This is what condensing boilers do by passing the flue gases through a heat exchange system which condenses the water that then runs out through an effluent pipe, and gives the heat to the return water flowing into the boiler. Dissolved carbon dioxide and some sulphur oxides in the condensed water make it mildly acidic with a pH between three and four. Because of this it should not be allowed to come into contact with metal that is susceptible to acid attack. The discharge piping is made of plastic and the heat exchanger of an acid resistant or appropriately coated metal.

Conventional boilers have efficiencies in the range of 60 to 85%. Condensing boilers have been available for some years with efficiencies over 85%. Their availability has allowed the legislation to be tightened again in the UK. Since June 2005 it has been required that all newly installed or replacement gas boilers be condensing where feasible. The same requirement applied to oil fired boilers from the 1st April 2007. In the case of coal fired boilers the fuel mainly consists of carbon rather than hydrocarbons so there is much less water vapour in the exhaust and condensing systems would carry little advantage.

The operation of condensing boilers is slightly different from conventional boilers. Conventional boilers raise the temperature of water flowing from them to between 75 and 85°C. The corresponding return temperature is between 65 and 75°C. In order to maximise the amount of exhaust gas water vapour that is condensed the return temperature has to be lower than this at between 50 and 60°C for condensing boilers. If the same 10°C difference between flow and return is used flow tempers will be in the 60 to 70°C range. This is somewhat lower than the corresponding temperatures for conventional boilers and because of this the heat output from radiators will be correspondingly lower. In principle radiators should be oversized to counter this but in practice radiator size and room matching is normally very inexact and errs considerably on the generous side so when a conventional boiler is changed for a condensing boiler it is rarely necessary to make any changes to the radiators. Condensing boilers will operate at high temperatures but with reduced recovery of heat from the exhaust gas water vapour. When this happens they are said to be operating in non condensing mode and efficiency suffers though remains higher than for conventional boilers. If the boiler is overpowered for the building the temperature drop from flow to return will be less than 10°C so the return temperatures will be higher favouring non condensing operation. This is a reason to avoid fitting an oversized condensing boiler.

Fuel

The most convenient fuel, and frequently the cheapest, is mains piped gas. It is almost always selected where it is available, its main limitation being that it is not universally available. Piped supplies have been installed in most urban environments but in more rural parts of the UK and large parts of other countries there is no mains gas supply.

Where no gas is available the usual choice is oil. The costs of heating with oil and gas are comparable but oil is subject to more variation in price than gas. Other disadvantages are that it has to be contained in a tank which is expensive to install, may be unsightly, and requires maintenance. The unsightly appearance of a tank is often worsened because they have to be mounted above the level of the boiler if gravity feed is to be used and this tends to put them in a prominent position.

Modern oil refilling services have become very sophisticated. It is now commonplace to have automatic level detectors in tanks that send a signal to the supply company when the tank is nearing empty. The supply company can then send a bowser out to fill the tank and the first to the owner knows of it is when they received the bill!

In the UK the market for oil fired boilers is smaller than it is for gas-fired and the range of options available is more restricted. The common types such as combination, standard, and cooking range are all available. The apparatus necessary to burn oil is somewhat more complicated than for gas and oil fired boilers tend to be slightly more expensive than their gas-fired counterparts.

Solid fuel

Central heating boilers that burn solid fuel are available but are rarely chosen because the fuel is dirty and inconvenient. There are certain circumstances where solid fuel fired heating might be chosen such as where a plentiful supply of wood is available but even then other heating options such as masonry heaters will usually be more competitive.

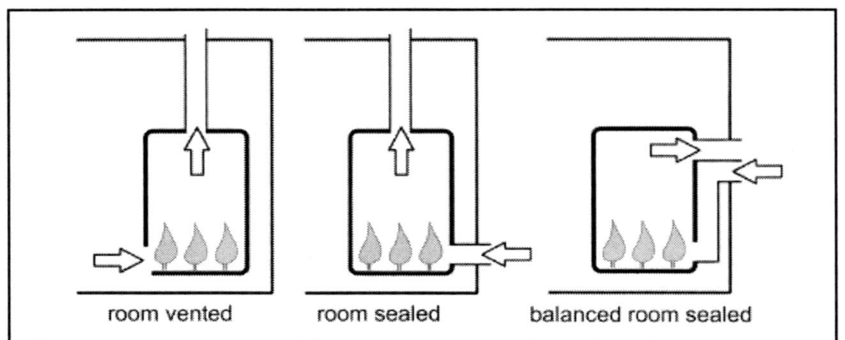

Figure 11.1: Diagram of flueing arrangements for boilers. The open or room vented system draws the air for combustion into the boiler from the room in which it is located. The room sealed system draws air from outside the room. The preferred option now is a balanced flue, a type of room sealed system where the air intake and exhaust outlet are at a similar height.

Bottled gas

In areas where there is no gas supply it is possible to install a central heating system that works on bottled gas. In the UK this is usually referred to as LPG (Liquid Propane Gas) which is a mixture of butane and propane. LPG is stored as a liquid in pressurised tanks. It boils at a low temperature and the vapour is carried off to fire appliances. It is a competitor to oil fired systems but is rarely chosen to fire central heating because it is considerably more expensive than oil, though this has not always been the case. The big advantage that LPG has over oil is that you can have a gas-fired cooking stove. Some people have a strong preference for gas-fired cooking over electric and where no mains gas is available that would justify the installation of an LPG system. In a building which is only intermittently occupied, such as a holiday cottage, an LPG fired central heating system would be a reasonable option because there would be no necessity for the installation of additional oil storage tank and piping facilities and the higher cost of heating could be tolerated if it was only used for short periods, especially if it was mainly used during the summer. A further advantage of LPG is that it provides the most satisfactory and cost-effective form of lighting available where there is no mains electric power. LPG fired gas lighting technology owes its advanced state of development principally to the boating and camping markets but it is occasionally useful in remote buildings.

Boiler resistance

The resistance of a boiler refers to the resistance offered to water flow through the heat exchanger. Virtually all modern oil and gas fired boilers are high resistance because efficient heat transfer is favoured by narrow water channels. Earlier cast iron water jacket boilers had low resistance. This has some bearing on central heating layout. Some layouts that work with a low resistance boiler do not work properly with a high resistance one. When changing an old low resistance boiler for a new one check the "T" rules of chapter 2.

Flues

Broadly there are two types of flue known as open or room vented and room sealed (figure 11.1). Room vented flues draw the air required to burn the fuel from the room in which the boiler is mounted and then discharged the exhaust gases to the outside, usually via a vertical flue going through the roof. This system has a number of disadvantages. Generous amounts of ventilation must be provided in the room which makes it cold. There is a risk that the boiler fire will be starved of air in which case it will produce the life-threatening gas carbon monoxide. These systems often employ flue piping outside of the boiler casing within the room that are prone to developing leaks allowing flue gases to enter the room. A combination of leaking flue pipe and air starvation allows carbon monoxide to build up in the room to lethal levels. For these reasons room vented systems are all but obsolete and should only be installed when a room sealed system is impractical, usually because the boiler in an existing installation is being replaced, its position is far from an outside wall and an existing room vented flue is already in place.

With a room sealed flue the air to burn the fuel is drawn in from the outside, passes through the fire and the exhaust gases pass to the outside with no air drawn from the room. Most boilers available now are room sealed and these are preferred. Room sealed systems where the air intake and flue outlet are close together are called balanced. The most common arrangement is to have a central pipe out of which the flue gases are discharged, surrounded by a wider pipe with the air being drawn in through the gap between the two. The system has the advantage that if a leak does arise in the flue pipe, the exhaust gases escape into the incoming air stream which is directed into the boiler rather than entering the room. Wherever possible this is the system of choice.

Flue siting is a complex matter. It is necessary to avoid flue gasses entering the house or surrounding buildings via windows or ventilators. Flue operation can be disturbed by strong winds or eddy effects around roofs. Structural and aesthetic considerations also arise particularly for chimneys above the roof line. If flue siting has to be chosen then the current building regulations should be consulted but this will usually be done by the CORGI or OFTEC (chapter 7) registered engineer who installs the boiler.

A leaking flue is potentially dangerous. Any soot staining seen around joints in the flue suggests leaking and should be investigated and repaired urgently.

Types of boiler

Standard boilers are the simplest containing only the necessary components to heat water. System boilers contain other components such as a pump, expansion chamber and control circuitry which simplifies the installation of systems. Like system boilers, combination or combi boilers also have pumps and expansion chambers within their casing. They also combine the functions of the space heating and hot water systems, hence their name. This simplifies system installation even further as there is no need to install a hot water system with cylinder, cistern and primary circuit.

Standard boilers

Standard boilers heat water but do not pump it or decide when it needs to be heated. This is a slight simplification as standard boilers have an internal thermostat that switches them off when the water temperature reaches a certain point but in normal operation this thermostat is usually not used.

Until fairly recently standard boilers were the only type available. There are literally hundreds of designs that have been produced over the years, many of which may still be encountered. Ironically the design life and durability of some of the earliest boilers was much longer than of more recent types which means that examples from virtually the entire history of central heating are in current use. These include solid fuel fired "natural draft" designs where air is drawn through the fire by the buoyancy of rising hot gasses in a wide chimney, oil fired natural draft designs, balanced flue designs, fan assisted designs, and contemporary condensing types.

System boilers

There is an increasing trend to combine different parts of the central heating system within the same casing as the boiler. These products are known as system boilers. They usually contain the boiler, safety devices, controls, pumps, and when designed for sealed systems they may also contain an expansion chamber and low pressure cut of sensors. Sealed systems usually operate at pressure of about one bar when cold which rises due to water expansion to between 2 and 2½ bar when the system is hot.

Combination boilers

Combi boilers are so called because they combine two functions. These are heating the water for the space heating and also heating water for the hot taps. A standard boiler also provides heat for the space heating and hot taps but a combi boiler does it without the use of a hot water storage tank. It does this by heating the hot water as it is required. This involves a complicated control system that senses when a hot tap is turned on, fires the boiler and moves motorised valves that divert mains cold water through a heat exchanger and into the hot tap circuit. When the hot tap is turned off this is detected and the motorised valves are moved back so that water from the space heating system is diverted to the boiler and the fire extinguished unless the space heating control system is calling for heat. They are always hot water priority.

Combi boilers bring a number of advantages. Unlimited hot water is available and this avoids the annoyance of drawing a bath and running out of hot water before it is full. No hot water cylinder is used and this saves space and eliminates heat loss from its surface which improves efficiency. Also the avoidance of a hot water tank generally reduces installation costs by more than the now quite modest price premium of combination boilers.

They also have disadvantages. They are very complicated and earlier models were prone to break downs though this has improved recently as their increasing popularity and market share has brought rapid development. To make the reliability problem worse there is no easy way of providing contingency against boiler failure comparable with the immersion heaters of hot cylinder and thermal store systems.

The power required adequately to heat hot water for the hot taps is usually quite a lot more than that required for the space heating system. The boiler may be too large for the house if adequate hot water flow is a deciding factor in its choice. If a condensing boiler is significantly oversized for the house it will tend to operate in non condensing mode with reduced efficiency.

A further problem with combi boilers is that the hot water temperature takes some time to stabilise. It typically goes through phases of being cold to being too hot before reaching a safe and comfortable temperature.

Finally the flow rate through combination boilers is limited. Maximum power output ranges from 24 to 40kW. The most powerful ones will raise the temperature of water by 35°C at a flow rate of about 16 l/min. This is close to the maximum flow rate available of cold water from the kitchen tap in many houses. The measurement is however somewhat misleading. The inflow of cold water in most cases has a temperature of between 4 and 14 °C. To achieve a hot tap water temperature equivalent to that of a storage tank system the hot water has to leave the boiler at a temperature close to 60°C, which typically means a 50°C temperature rise. This reduces the corresponding flow rate quite substantially to 11.2l/min. The rate at which hot water can be supplied by the less powerful models is correspondingly lower. The most powerful combination boilers of 40 kW are able to provide hot water at an adequate flow rate for two taps at once or a good flow rate for one.

These disadvantages can be overcome by the use of a thermal store hot water system that is heated by a combi boiler. These systems are discussed in chapter 3.

Combi units

In essence a combi unit is the same as a combination boiler. The difference is that a combi unit contains a small hot water storage tank within the boiler casing which contains a few litres of water that is kept hot all the time. This largely gets round the problem of temperature instability when a hot tap is turned on and is a significant advantage. The disadvantage is that it makes the boiler casing somewhat larger but if the size can be tolerated a combi unit is an excellent choice.

Boiler position

One decision to make when installing a boiler is whether it should be wall or floor mounted. Early central heating boilers used cast-iron water jackets surrounding fairly large combustion chambers. Their large size and great weight meant that they had to stand on the floor. As boilers have become more sophisticated they have become smaller and lighter making it possible to mount them on a wall, saving floor space, and the majority of boilers installed today are gas-fired models of this type. Many floor standing boilers are manufactured and they tend to be the larger more powerful type. There is also a tendency for oil fired boilers to be floor mounted for the simple reason that the most reliable and inexpensive means of driving oil from the tank to the boiler is gravity feed, a system that is enhanced of the border is at low-level.

Impinging on where to position a boiler is the question of what kind of flue should be used. Most commonly a balanced flue will be used in which case the boiler is best mounted on an outside wall through which the flue can pass. If another type of flue is to be used its positioning is likely to determine the boiler position rather than the other way round.

Boiler components and faults

Boiler thermostats

Boilers are equipped with internal thermostats. There is always an overheat thermostat. It is placed in the heat exchanger and shuts down the boiler if the temperature becomes to high. This thermostat will only be activated if there is a malfunction such as too little water movement or a dry fire. A dry fire is when the water jacket or heat exchanger are not filled with water. Under these circumstances the flame temperature can rapidly melt the metal from which they are made. The overheat thermostat is a failsafe mechanism that is not activated in normal operation. It should not be tampered with!

As well as an overheat thermostat, most boilers also have a boiler thermostat. This is intended to control the maximum water temperature or the difference between the flow and return temperatures during normal operation. In the past the boiler was controlled by a time switch only. Hot water was pumped or convected around the space heating and primary hot water circuits according to demand. If there was no demand for heat the boiler heated the water in its restricted circuit until the boiler thermostat switched it off. The water then cooled until the thermostat switched it back on again. This is called boiler short cycling and is wasteful of energy. It is prevented if the boiler is switched off by the control system when heat is not being called for. This control feature is called boiler interlock. In the UK the building regulations now required that boiler interlock is installed. In modern well-designed systems operating normally the boiler thermostat is not used.

Fuel control valve

The valve system that controls the flow of fuel to the boiler is a complicated thing with various solenoid and failsafe features. Perhaps not surprisingly it is failure prone. It is the preserve of qualified engineers. Its complexity makes it expensive.

Flue fan and pressure switch

Most modern designs of boiler use a fan to drive air into the fire or to draw flue gases out of it. This allows for the boiler to be more compact and enhances the burn speed possible with balance flue systems. The problem is that if the fan should fail the fire will be starved of air and flue gases may accumulate. If there is a leak in the flue pipe flue gas may escape into the room. To avoid this a further failsafe system is necessary. This is a sensor device that detects when the fan is operating. It uses a pressure switch that detects the pressure change which the fan would normally produced. Fans are rather failure prone components and when they fail the pressure switch ensures the boiler will not start. Occasionally the pressure switch itself may fail and in that event the fan will turn on but the boiler won't fire and the fan will quickly turn off again.

Boiler kettling

Modern boilers are not intended to boil water. If the system is operating normally water is pumped through the heat exchanger fast enough that the temperature never rises to near boiling. If something is going wrong and the water is flowing to slowly because the pump is failing, has too low a speed setting or some blockage is occurring then water in the heat exchanger will be heated to a higher temperature than normal. The boiler's internal thermostat will cut the boiler off before it reaches 100°C but the temperature in very small areas very close to heat exchange surface may do so giving rise to kettling. Kettling is the formation of microscopic steam bubbles on the surface of the heat exchanger which then rapidly collapsed because the water immediately beyond the heat exchange surface is cooler than boiling point. Kettling gives rise to a sound which is familiar because it is the same sound produced by an electric kettle as it heats the water. Such a sound coming from a boiler indicates a low flow state and should be investigated.

CHAPTER 12
Cisterns

The terms "cistern" and "tank" are often used interchangeably. Here I will stick to the term cistern for the float valve operated cold water containing vessels that are usually located in the roof space. This is to avoid confusion between cisterns and hot water cylinders that are also referred to as tanks.

Until fairly recently cisterns were standard components and still predominate in existing installations. They have a number of disadvantages. They carry the risk of legionella growth and burst pipes due to frost can cause major water damage. Running pipe work to the attic is expensive and the large weight of water that must be stored leads to structural requirements. They can take up considerable attic space and oxygen may dissolve in the central heating water leading to oxidative corrosion.

When a new water or heating system is being installed, cisterns usually can be and should be avoided. Most modern cold water supplies are consistent and reliable meaning that cold water taps can all be connected to the rising main with a suitable non return valve to prevent contamination of the main supply. This avoids a cold water cistern. Hot water systems using modern combi boilers, thermal store tanks, or un-vented cylinders do not require a cistern and where possible one of these options should be chosen. Modern sealed central heating systems likewise do not require an expansion cistern.

Cisterns are not without their advantages. Systems using them run at low pressure which tends to minimise stress on components and means that when leaks do occurred they are not too fast. They allow high water flow rates at the taps even if the mains supply is slow or intermittent and they provide additional protection against backflow into the mains supply.

It is usually desirable to have the highest feasible water pressure for hot and cold taps and for radiators the pressure should be above the top of the highest radiator in the house. For these reasons cisterns are usually located in the roof space. Because of the risks the installation and maintenance of cisterns are subject to close regulation.

Structural requirements

Cisterns typically contain hundreds of litres of water so must be on a strong support. They should be placed on beams with not less than 350 mm between centres. Galvanised steel cisterns are sufficiently rigid to rest directly on the supporting beams but nowadays most are made of plastic which is more flexible. Plastic cisterns must be placed on timber boards or a piece of marine plywood that is at least 25mm thick and is large enough to support the whole of the bottom of the cistern. Marine plywood is used because it must withstand getting wet. Chipboard should not be used because if it gets wet it becomes structurally weaker and disintegrates. All components of the tank should be able to tolerate boiling water. It is good practice to provide some form of waterproof drip tray so that any leakage from the tank does not soak through the ceiling.

Plumbing requirements

Regulations control the positioning of the various pipes connected to cisterns. The principal risks to avoid are back-flow of water into the mains supply which must not be allowed to happen even if the mains supply pressure at the cistern inlet becomes negative. To prevent this it must be ensured that the outlet from the float valve is never submerged. This requirement is not met by some valve extension pipes which are intended to reduce noise by delivering water under the surface. The rules in the UK for domestic cold water cisterns under 1000l in capacity are shown in figure 12.1.

The float valve is usually placed about 50mm down from the top edge of the cistern. With plastic cisterns the walls are fairly flexible and a float valve attached to them would be able to move so a rigid backing plate must be used. The level of water at which the float valve closes is adjusted with a screw.

There must be an overflow pipe fitted and it must have a diameter larger than the inlet pipe. It must be placed so that its bottom is not less than 25mm above the water level at which the float valve closes. Its end may be dip into the water as shown in figure 12.1 to prevent cold air entering the cistern and promoting freezing in which case an air vent must be provided to prevent siphoning. The vent must be covered by a mesh with hole size not more than 0.65mm.

The outflow pipe should have a service valve fitted that should be below the insulation but its handle should protrude above. The outflow should be taken from the side opposite the inflow to prevent water in the far end from the inflow stagnating as can happen with long low "coffin" type cisterns.

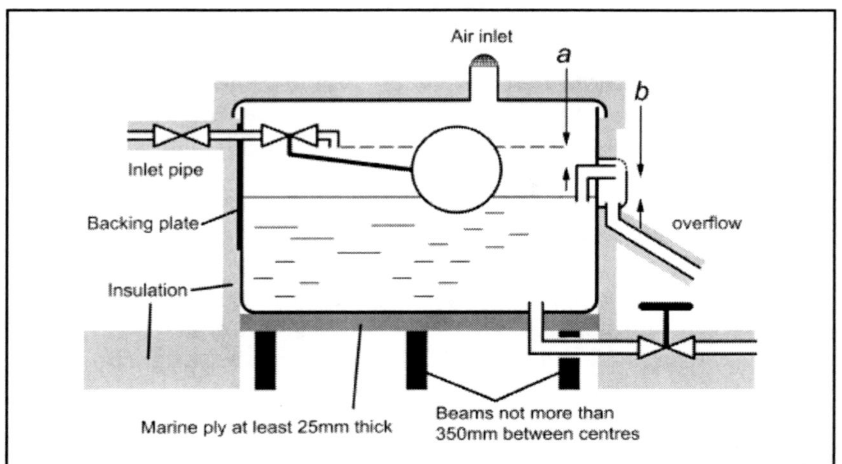

Figure 12.1: Diagram of a cold water storage cistern. The inlet pipe should be at least 15mm in diameter, be well insulated and have a servicing valve fitted. If a plastic cistern is being used fit a rigid backing plate behind the float valve outside the cistern to prevent flexing of the cistern wall altering the water level at which the valve closes. The air inlet should be covered by a corrosion resistant grill with a hole size not more than 0.65mm to exclude insects. The inlet opening should be not less than 25mm or twice the diameter of the inlet above the top of the overflow (a) to prevent backflow into the mains. The bottom of the overflow should be not less than 25mm above the level at which the float valve turns off (b). The overflow may dip below the surface to prevent cold air entering and promoting freezing. If it is dipped there must be an air vent as shown to prevent siphoning and it must be covered by a mesh to exclude insects.

It is a legal requirement in the UK to inform the local water authority if any works are being undertaken that lead to a potential risk of contamination of the main water supply. Five days notice must be given for such works.

Insulation

Cisterns and all associated pipe work should be insulated to prevent freezing and to keep the temperature under 20°C to prevent legionella growth. The underneath of cistern should only be insulated if it is installed in a building that is not heated.

Cold water cistern

A cold water storage cistern is used to store water for the cold taps. They must mow be installed to a potable standard – meaning that the water can be drunk. Special paints can be used to coat the inside of old cisterns to bring them to this standard. Even so where a cold water cistern is in use the kitchen sink tap must not be supplied from it but directly from the rising main to provide a source of wholesome drinking water in the event of cistern contamination. They are used where the mains supply is intermittent or slow. They may be used if the mains has low pressure in which case a pump may be needed to raise the water to the roof space. They are also used when a building is not on the mains water supply and draws water from a well. The size of cistern depends on circumstances. It should contain 12 - 24 hours supply of water. In most cases 115 l plus 100 l per bedroom in the dwelling is about right.

The cistern should have a well fitting rigid but not airtight lid. In this there should be an air inlet covered by a wire mesh to exclude insects. The mesh should be corrosion resistant and have a maximum holes size not more than 0.65mm.

The float valve outlet spout must not be less than 25mm or twice the spout diameter whichever is greater above the top of the overflow. The bottom of the overflow must be not less than 25mm above the highest water level during normal operation. Inlet and outlet shall have service valves.

If a cold water cistern is being used it can double as the hot water cistern in which case its capacity should be raised by 50%. The hot water outlet should be above the level of the cold water outlet and both should have a servicing valve. The hot water vent pipe if used should enter the lid through a specially made cylindrical sleeve and terminate not less than twice its diameter above the water level when the cistern is full.

Work done on the cold water supply system is liable to dislodge debris and send it into the cold water circuit where it can block or jam taps and valves. Ceramic taps and the float controlled valves in toilet cisterns are the most susceptible. If possible these should be isolated while work is done. After completion of the work run a tap fully open for a minute or two to flush debris out and check the operation of all toilet flushes. If a ceramic tap is affected by debris it usually drips because the disks are held apart. If this happens open the tap fully to clear the debris. Trying to stop the drip by forcing the tap closed risks breaking the disks.

Hot water cistern

A hot water cistern contains cold water but to supply the hot water system. If it is not also serving as a cold water cistern its capacity should be about 45l / person living in the dwelling. The plumbing details are as for the cold water cisterns and one cistern can be used for both.

Central heating expansion cistern

Unlike other types of cistern, central heating expansion cisterns must accommodate thermal expansion of water as well as top up the water to replace loses. System water typically expands by 2% from cold to hot. 4% is possible in extreme cases with malfunction. To provide a margin of safety 5% should be accommodated.

The outflow pipe must be at least 15mm in diameter and must only connect to the central heating circuit and nothing else. It must contain no valves.

The vent pipe should be at least 22mm in diameter, should end below the lid of the cistern and not less than twice its internal diameter above the top of the float valve. It should rise continuously with no reversals from its point of connection and must contain no valves.

The overflow pipe should be high enough that the bottom of its opening in the cistern is not less than 25mm above the water level at full operating temperature. Recommended cistern and pipe sizes are given in table 12.1.

The float valve is adjusted so that the water level when the system is cold is 1/3 of the way up the tank to the overflow or 100mm above the bottom of the tank which ever is greater.

Adding chemicals to the feed and expansion cistern

Chemicals such as cleaning agents, descaling agents or inhibitors can conveniently be added to an open vented central heating system via the cistern. Make sure it is the correct cistern! The central heating cistern is small and may be inconspicuous. The first step in adding chemicals is to tie up the float valve. Note the volume of chemical solution that is to be added and drain this much water off from a low drain point in the central heating circuit, then pour the chemical solution into the cistern. Leave the float valve tied up. To circulate the water round the system

Table 12.1 Central heating feed and expansion cistern sizes

Boiler power (kW)	Cistern size (l) ≥5% of system volume	Feed pipe diameter (mm)	Overflow pipe diameter (mm)	Vent pipe diameter (mm)
Under 25	45	15	22	22
25-45	70	22	28	28
45-60	90	22	28	35

drain off about 5l of water from a low drain and pour it into the cistern. Repeat this until the total volume drained off and poured back is equal to twice the cistern volume.

Problems with cisterns

Overflow

Water overflowing from cisterns should be easily noticed because overflow pipes are deliberately located in conspicuous places so that overflow can be recognised. Cisterns overflow because of failure of the float valve to close properly. The commonest reason is that grit carried in the incoming water has caught in the valve preventing it from closing. This usually clears when the cistern empties and refills. Other common faults are that the closure mechanism is stiff or the valve arm is bent.

Empty cistern

The commonest cause for an empty cistern is that the float valve is stuck in the closed position. This may be because the mechanism is stiff or seized or because the float arm is bent. If either a hot or cold water storage cistern is empty there is no water flow from the relevant taps. If the central heating cistern is empty the fault will be more subtle as the central heating will still be mostly full of water so will work. Air collects in the top of radiators. The pressure in the system falls as the water level falls down the cold feed pipe so the air cannot be bled out of the tops of radiators, particularly on the upper floor.

Outflow blockage

Cistern outflow blockage is caused by corrosion products, debris that falls in, ice or biological slime. If blockage occurs the cause should be found and corrected as well unblocking the outflow. Any signs of biofilm development should be actively looked for. If the central heating feed and expansion cistern is blocked check the inhibitor level.

The symptoms of blockage vary from system to system. A blocked outflow from a cold water cistern results in no water or slow flow from the cold taps. With a hot water cistern outflow blockage, rather than flow being particularly slow the hot taps splutter because air is drawn down the cylinder vent pipe and into the secondary hot water circuit. Blockage of the outflow from the central heating feed and expansion cistern results in the same symptoms as an empty cistern; air collects in radiators and cannot be bled out, particularly from radiators on upper floors.

Biological contamination

Biological contamination of any cistern is a serious development and should be urgently investigated and corrected. The principal concern is that whatever its nature, biological contamination indicates a significant risk of legionella growth. Antimicrobial chemical products are available to flush contaminated systems before they are put back in service but it is necessary to establish why the contamination occurred and prevent further occurrence. Check that the cistern is properly covered, clean and suitably positioned to maintain a temperature under 20°C. Check that the cistern is suitably sized for the building and that the outflow is on the opposite side from the float valve so that there is regular flow of water throughout the whole cistern and no areas are left stagnant.

CHAPTER 13
Control systems

Central heating control systems are designed to achieve to two things: automatically turn the heating on or off at appropriate times and while the heating is on maintain a constant comfortable room temperature and suitable water temperature. The standard control system uses hardwired mains voltage cabling that both powers and controls the boiler, pump and motorised valves. Increasingly popular now are more versatile systems that use either hardwired low voltage or radio signalling from thermostats to a central microprocessor controlled programmer that powers and controls the boiler, pump, and valves. Several manufacturers make systems and they are simple to assemble and come with comprehensive instructions so that little knowledge of wiring is necessary for their installation. With these fault tracing is a matter of deciding whether individual units are dead or alive rather than tracing through potentially poorly installed wiring. Wiring up conventional mains voltage systems is comparatively complex for which reason most of this chapter and chapter 13 are devoted to the subject.

Temperature control

The commonly used type of temperature control system employs a thermostat. This is a simple device with a temperature sensitive component such as a bimetallic strip or fluid filled expansion chamber that controls a switch. The switch is on at low temperatures and off at high temperatures. The temperature at which the switch turns on can be adjusted. When the temperature is low the switch turns the boiler on. The temperature then rises until the thermostat switches it off at which point the temperature starts to fall until the cycle repeats. The temperatures at which the thermostat switches on and off are different, the switch on temperature being the lower. The difference between these temperatures is called the temperature differential of the thermostat. The speed of the cycle depends on the rate of heat provision and loss and on the temperature differential. Rapid cycling carries a cost in terms of ware on control components and a small penalty in efficiency. How stable the temperature must be varies with application and where less stability is required thermostats have wider differentials. Tank thermostats that control the temperature of hot water cylinders have fairly wide temperature differentials of 5 to 10°C. Room thermostats have much narrower differentials, Low limit return pipe thermostats have the widest differentials of 20°C or so.

Room thermostats

The temperature that is comfortable for most people in a room is between 18 and 22°C. Gradual variations in temperature within this range of up to about 3°C are generally not noticed by room occupants but rapid changes may be. Room thermostats are designed to control the temperature between narrow limits and have a low temperature differential, 0.6°C being typical for modern units.

Figure 13.1 is a circuit diagram of a room thermostat. Not all thermostats have all four connections shown in this diagram. At its simplest the thermostat consists of a switch between pins 1 and 3. This switch is on when the thermostat is cold and off when it is hot. Pin 4 is provided in some thermostats for a "hot load" which might be an air

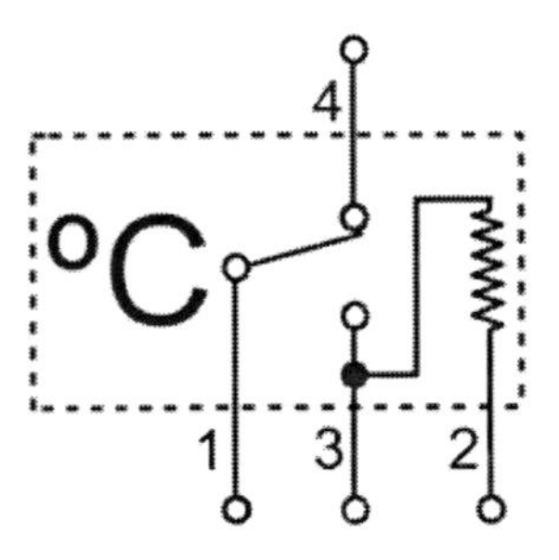

Figure 13.1: Diagram of a room thermostat with an accelerator resistor. Pin 1 is the switch common. Pin 3 is normally connected and is not connected when the thermostat is above its set temperature. Pin 4 which may or may not be present is normally not connected and connected when the thermostat is above its set temperature. Pin 2 is connected to Neutral if the accelerator is to be used.

conditioning system for example. This is on when the thermostat is hot and off when it is cold.

A problem with thermostat controlled central heating systems is that when the thermostat switches the boiler off, the temperature of the room continues to rise because the radiators take some time to cool down. This means that the room temperature is inclined to overshoot the thermostat "switch off" temperature which can lead to wider variations than 3°C. This problem is at its worse when the heat capacity of the radiators is high. This tends to be when old-fashioned cast-iron radiators or other high water content radiators are in use. Non-convector radiators contain about twice the amount of water as convector radiators and have a higher heat capacity for equivalent output powers. Thermostats intended for use with high water content radiators include an electric heating element called an "accelerator" or "anticipator". This is a resistor which warms up when the thermostat is calling for heat. It is connected between pins 2 and 3. This warms the thermostat slightly above room temperature so that it switches off before the room reaches the preset temperature. This has the effect of reducing the preset temperature only when the thermostat is calling for heat, i.e. when heating is switched on. It compensates for the effect of high heat capacity radiators by switching the heating off early when the temperature is rising because the heating is on rather than for some other reason such as that the sun is out or a fire is lit.

The same effect can be achieved more accurately by modern digital systems that measure the room temperature and learn how it responds to heating rather than simply switching at a preset temperature.

The naming of pins 1 to 4 in figure 13.1 is not universal. Some manufacturers designate pin 1 as "C" for common. Check the diagrams that come with the thermostat.

The optimal location for a thermostat is 1½ metres above floor level on an internal wall where the temperature is representative of the whole zone. It should not be placed behind a curtain or directly over a radiator! The simple system of using a single thermostat to control an entire house, or even an entire floor, has the problem that if the thermostat is installed in a location that is not representative, the rest of the building will be too hot or too cold. A popular location for a thermostat is the hall and it is a reasonable choice but if the front door is left open frequently then the hall is likely to be cooler than the rest of the house. The thermostat will call for heat appropriate to warm the hall but that will make the rest of the house too hot. If a thermostat is installed in a location where there is an additional heat source apart from the central heating system such as perhaps a gas fire or open fire, then the secondary heat source may adequately heat the room, keeping the thermostat off irrespective of how cold the rest of the house gets. For this reason a thermostat controlling several rooms should not be installed in a room where there is a frequently used secondary heat source.

An improvement on using a single thermostat is to use heat zones, each of which has its own thermostat. This requires a more sophisticated control system because it must not only turn the boiler on or off but must also control motorised valves that divert hot water to the zone calling for heat.

Tank thermostats

Tank thermostats are designed to control the temperature of water in the hot water cylinder. They have a dial marked in °C that can be set manually to the temperature at which they switch on. They are factory preset at 60°C and in general this temperature should be used. The exact temperature of hot water in the cylinder is not that critical for which reason Tank thermostats have fairly wide temperature differentials of between 5 and 10°C.

Figure 13.2 is a circuit diagram of a tank thermostat. Unlike the room thermostats of figure 13.1 there is no variability in the number of pins. Tank thermostats always have three pins because both on when cold and on when hot pins are necessary for some control circuits as explained in chapter 14. The labelling of the terminals varies from manufacturer to manufacturer. Honeywell call terminal 1 in this diagram "C" for common, terminal 3 which is on when cold and carries

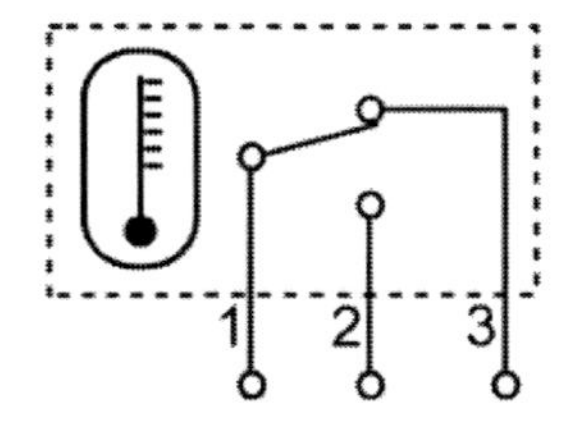

Figure 13.2: Diagram of a hot water cylinder thermostat (tank thermostat). Pin 1 is the switch common. Pin 3 is normally connected and is not connected when the thermostat is above its set temperature. Pin 2 is normally not connected and is connected when the thermostat is above its set temperature.

current to the heating load they call 1, and 2 which is on when hot and carries current to the cooling load they also call 2.

Frost thermostats

If the water in a central heating system is allowed to freeze, extensive damage can result. This takes the form of split pipes or radiators, damaged control valves and damage to the heat exchanger or water jacket of the boiler. There is a risk that this will happen if the house is left unoccupied with the heating switched off for several days in the winter. Over this time period the temperature of the fabric of the building slowly falls to that of the outside and can easily fall below freezing point in cold weather. The additives used in central heating water act as a kind of antifreeze reducing the freezing point of the water but it is still quite possible for it to freeze. A frost thermostat is intended to prevent this from happening. It is a low temperature thermostat that is wired to override the other control systems except for the mains double pole switch used to shut down the central heating system for maintenance and repair. It is wired to override the programmer and ideally the room thermostats as well but as most room thermostats are manufactured with a minimum temperature setting of 5°C or more, in circumstances where frost damage is likely they will probably be switched on anyway so this is less critical.

Frost thermostats are specially made with a low temperature range of 0 to 20°C. The standard setting used is 5°C. Some frost thermostats are fixed at this temperature and cannot be adjusted. When installing a frost thermostat you must be careful to take precautions against it being mistaken for the ordinary thermostat. If it is mistaken and turned up to 20°C it will result in the building being heated all of the time by virtue of its being wired to override the programmer. They come with a tamperproof cover for this reason.

In order effectively to prevent freezing in any part of the central heating system the frost thermostat should be located in the coldest part of the house where central heating pipe work or other water containing components are located. If there is no radiator in this area the frost thermostat will activate the central heating system and continue to call for heat as little heat is delivered to the area where the thermostat is. A low limit pipe thermostat gets round this problem (see below) and should be fitted in these circumstances.

Many boilers now have built-in frost thermostats that can be relied on to protect the system in cases where the temperature of the whole house is expected to follow the outside temperature fairly uniformly. A risk of freezing will remain in cases where the boiler is installed in the centre of a house where the extremities cool faster and may drop below zero before the frost thermostat in the boiler is triggered. The reverse can happen if the boiler is located in an outbuilding or Garage. In this case the environment of the boiler itself is likely to drop below freezing more quickly than the house and if the boiler has no internal frost thermostat a frost thermostat should be installed next to or just above the boiler. A complication here is that there is unlikely to be a radiator in the outbuilding were the boiler is located. In order to prevent the boiler from firing continuously while the temperature in the outbuilding is below 5 °C an additional low limit pipe thermostat is used. This is a thermostat attached to the return pipe on its way into the boiler. It is wired in series with the frost thermostat and turns off the central heating system when the pipe temperature rises over 20 to 25°C. This means that the boiler is only fired when both the outside temperature is below 5°C and the return pipe temperature is below 25°C. This keeps the fuel used to protect from frost at an acceptably low level while giving guaranteed protection. Low limit pipe thermostats have a wide temperature differential of 20°C.

Programmers

The term "programmer" is now used to refer to the timing side of Central heating control systems. At their simplest they are electromechanical switches but these are now becoming less popular in new installations in favour of digital electronic models. These are reliable, inexpensive, and extremely flexible but programming can be complicated and unforgiving, and quite often the displays are difficult to see; important limitations in systems for the use of elderly or infirm people.

Simple electromechanical devices generally allow for the heating to be turned on and off either once or twice a day. They also allow for both the hot water and space heating to be turned on or off together, or for the hot water to be on while the space heating is off. They do not generally allow for the hot water be to be turned off while the space heating is on as this combination is unlikely to be required. Programmers that have these three options only are now called mini-programmers. Most modern programmers allow all four options and these are referred to as programmers rather than mini-programmers.

Modern programmers offer a variety of other features. Some require that the same time settings are used every day and are known as 24-hour programmers. Some allow two different timings one of which is used on Saturday and Sunday and the other is used Monday to Friday. These are referred to as 5/2 programmers. Still further ones are available that allow independent timing to be used for each day of the week and theses are referred to as 7-day programmers. As they have become more sophisticated the time and effort required to program them has increased considerably. Many electronic devices have "volatile" memory which is lost if the power is switched off. This can cause considerable irritation because the entire system has to be reprogrammed. To get round this the most up-to-date programmers have on board batteries which allow them to retain the programming information in the event of a mains failure.

Programmers come as two parts; one is a terminal block to which the wires to the boiler, pump, thermostats and motorised valves are connected. This part has cable connections on one side and a complex socket on the other into which the programmer proper plugs. Many modern programmers are designed as plug-in replacements for older versions allowing easy upgrading.

Low voltage and radio systems

Conventional mains voltage control systems require that the thermostats switch mains voltages with high enough currents to power the boiler, pumps and motorised valves. More modern systems are available that avoid this by using a microprocessor controlled central programmer that takes data input from thermostats and uses electronic switches to control the boiler, pumps and valves. The data input from thermostats can come in the form of low voltage hardwired connections or radio links. There are various restrictions connected with the radio frequency transmitted by the radio type which are best not installed near other equipment that uses radio waves and should not be mounted in metal wall boxes.

The thermostats of these systems can have several programmable functions that allow more flexibility and combine the function of a thermostat and time switch. They allowed different temperatures to be set at different time periods and may even have an optimisation feature where they can estimate when it is necessary to turn the heating on in order to achieve the target temperature by a certain time rather than simply switching the system on at the specified time. This is a particularly nice feature in the mornings in the winter as people can time the central heating system so they get up to a warm house rather than one that is cold though getting warmer.

These systems are simple to install requiring little knowledge of wiring and come with comprehensive instructions of how to connect them up.

Chapter 14
Wiring

For many years hardwired mains voltage systems have been the standard for central heating control but more modern systems using electronic programmers and low voltage or radio thermostats are replacing them. Electronic systems are more flexible and require little detailed knowledge to install correctly. Wiring and repairing mains systems demands a more detailed understanding of their function and is the subject of this chapter.

Wiring up complex central heating control systems involving several zones, room thermostats, tank thermostats, frost thermostats and possibly a low limit pipe thermostat, and an elaborate programmer can be quite a headache! The job has been considerably simplified by many manufacturers who make all the necessary components and generally use a common colour coding system for the cables necessary to join the bits together. Manufacturers like Honeywell provide comprehensive and relatively easily followed wiring diagrams that largely avoid the need for a detailed understanding of how the system works. That is fine if it works first time but if not or if it later fails a more detailed understanding will be necessary.

If you have not read the section on electrical safety in chapter 4, go back to it. If you install a system that doesn't work properly it is a bit embarrassing but can be sorted out. If someone gets electrocuted it is a different matter!

There follow circuit diagrams and explanations of control systems for various central heating layouts. These circuits are simplifications of the manufacturers wiring diagrams in two respects. Firstly wiring diagrams usually include the connection board which is typically part of the back panel socket of the programmer. The programmer plugs into it via a standard configuration of pins and sockets. This means that the programmer can be replaced or upgraded by simply un-plugging one and plugging another into the back panel. Most forms of programmer that have been used over the past 20 years can be replaced or upgraded. The industry is moving towards a common plug configuration for all manufacturers but this has not always been the case and manufacturer's literature should be consulted to determine which programmers can be upgraded to what.

The second simplification in the diagrams below is that the earth and neutral wires are not shown because connecting them up is simple and they make it more difficult to follow the circuit diagrams. Earths from all units are connected to an earth point and all neutrals to a neutral point. The difference is that the earth point is not switched whereas the neutral point is; the connection is on the central heating side of the double pole mains switch. These simplification are used because they make it easier to understand how the circuits work.

Colour coding

The standard colour coding of electrical wires is brown for live, blue for neutral, and green and yellow stripes for earth. When standardised colour coding was first introduced a more intuitive system of a red for live, black for neutral, and green for earth was used. This was easy to remember but unfortunately there was a problem. Red/green colour blindness is quite common among men. This had the rather unfortunate result that a significant number of men could not see the difference between earth and live! As a consequence the colour coding system had to be changed. After careful consideration of the various prevalent forms of colour blindness the brown for live, blue for neutral and green/yellow for earth system was introduced. The red green confusion problem was specific to electric flex and did not apply to cables in which the live and neutral wires are insulated but the earth is bare so colour blind people could tell the difference between live and earth. The red/black colour coding of live and neutral remained in use for cables but in the interests of consistency and simplicity the coding was changed to brown for live and blue for neutral in April 2006.

Motorised valves come connected to colour-coded cables. The colour coding is now fairly standardised across the industry. Blue is neutral and green/yellow is earth. Other colours used very with the type of valve. In diverter valves where there is only one live terminal it is coloured brown.

In 2 port and mid position valves there is more than one terminal which is potentially live. In two port valves only one wire activates the motor and it is brown.

In three port mid position valves the motor is activated by two wires, white and grey. If the white wire only is live the valve moves to the mid position. If both white and grey wires are live the valve will open port A and close port B. If 240V is supplied to the grey wire only the valve will be held in its last position.

Two port and mid position valves include an internal switch which indicates the valve's position. It works by connecting the incoming live wire to an outgoing wire. This outgoing wire is orange or red.

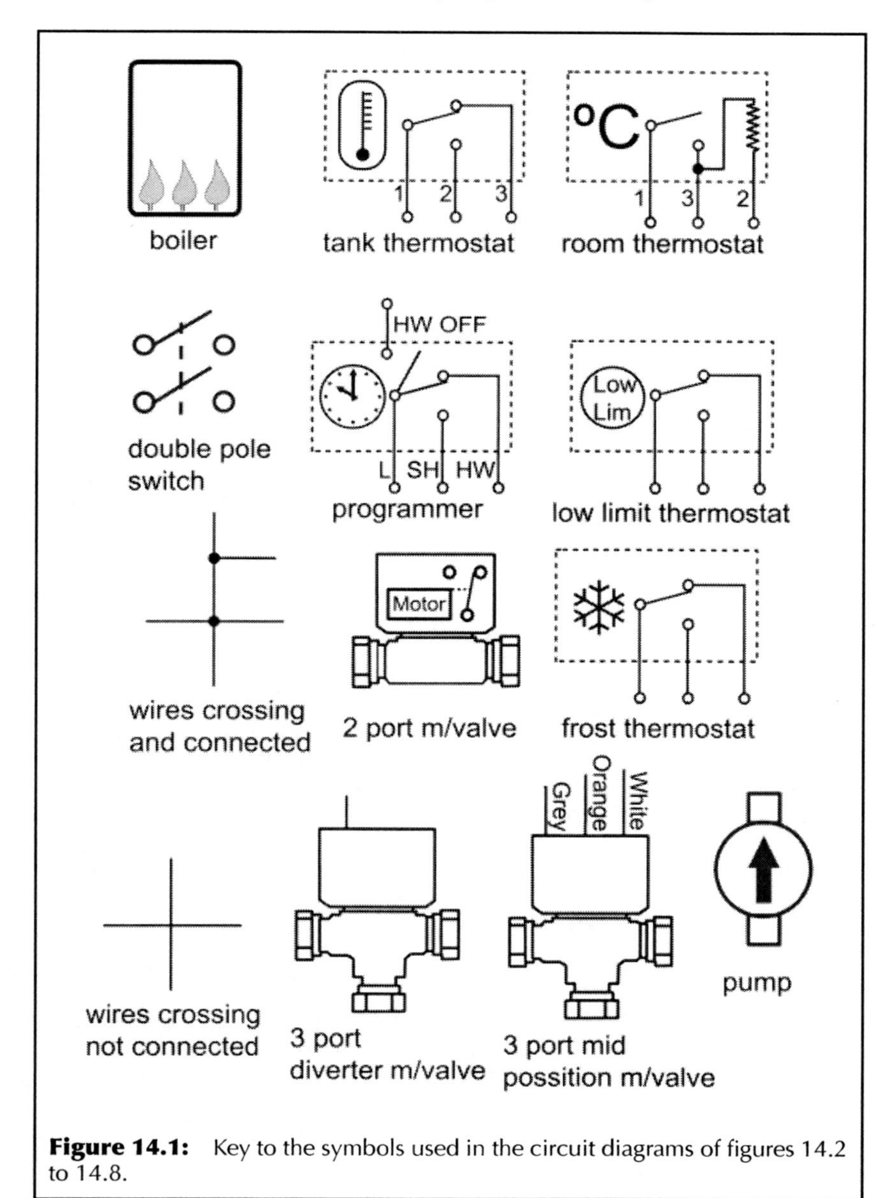

Figure 14.1: Key to the symbols used in the circuit diagrams of figures 14.2 to 14.8.

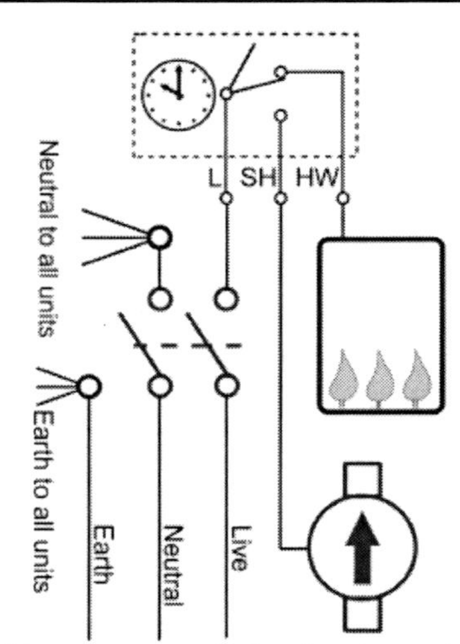

Figure 14.2: Central heating control circuit using only the boiler's internal thermostat for temperature control. This layout is intended for partly pumped systems and is now obsolete but many remain in service. They were used with time switches that allowed space heating and hot water to be on simultaneously, hot water to be on alone but not space heating alone. Time switches with these limitations are known nowadays as mini-programmers.

How it works

When the mini-programmer calls for hot water only the boiler is fired but the pump does not run. Hot water is driven via convective currents through the boiler and round the primary hot water circuit giving up heat to the water in the cylinder. As the hot water cylinder temperature rises the temperature of the water returning to the boiler rises. The flow temperature also rises until the boiler's internal thermostat turns it off. The temperature then slowly falls until the boiler's internal thermostat turns it on again and the cycle repeats. If hot water is drawn off from hot taps in the house cold water enters the cylinder from the bottom. This cools the water in the primary circuit promoting convective currents which drive water through the boiler's heat exchanger or water jacket. The boiler thermostat then turns it on and the boiler is fired until the cylinder is again filled with hot water.

When the time switch calls for both hot water and space heating the boiler is on and the pump is on. The pump drives water through both the space heating circuit and the primary hot water circuit. It is unlikely that the circulating water will reach a high enough temperature to prompt the boiler's internal thermostat to turn it off. In most systems of this type therefore space heating tends to be overprovided. Temperature control can be added to this system either by upgrading it to one of the circuits below or by adding thermostatic radiator valves.

The requirement that the pump should not be operating while the boiler is not is provided by the time switch which does not allow the space heating circuit to be live while the hot water circuit is not.

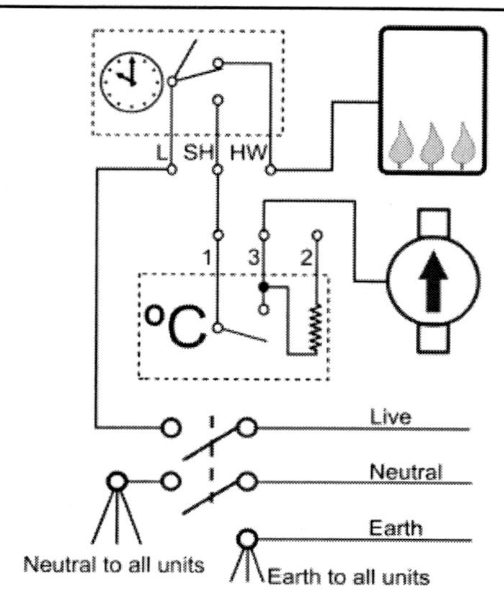

Figure 14.3: This circuit is now obsolete but as with the circuit of figure 14.2 many remain in service. It is the same as the circuit above except that a room thermostat has been added to the space heating circuit. Assuming that the thermostat is properly located this is the single most significant way in which inefficiency can be improved over that provided by the control system of figure 14.2.

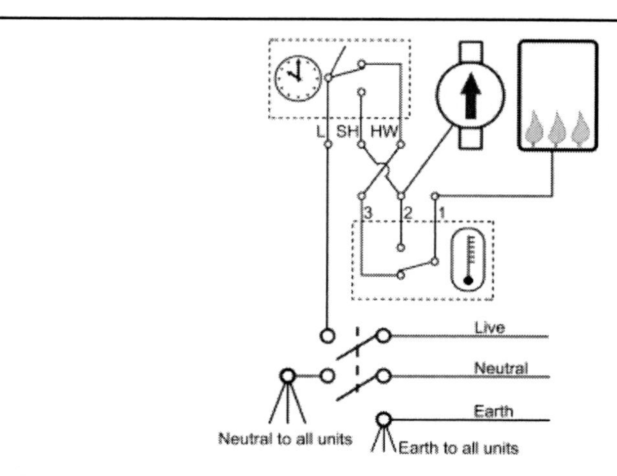

Figure 14.4: In this circuit a tank thermostat has been added to the circuit of figure 14.2. The changes are more complex than with the addition of a room thermostat. This circuit provides a form of boiler interlock, that is that the boiler is switched off by the control system when the temperature sensors are not calling for heat.

How it works
When the mini-programmer is switched to hot water only and the tank thermostat is calling for heat it connects pin 1 to pin 3. The boiler is thereby switched on. When the tank thermostat's call for heat is satisfied it switches pin 1 to pin 2. This has the effect of putting the boiler and pump in parallel so that they are switched on together if the mini-programmer is calling for space heat. This provides boiler interlock because when hot water is on but the thermostat is not calling for heat the boiler is switched off.

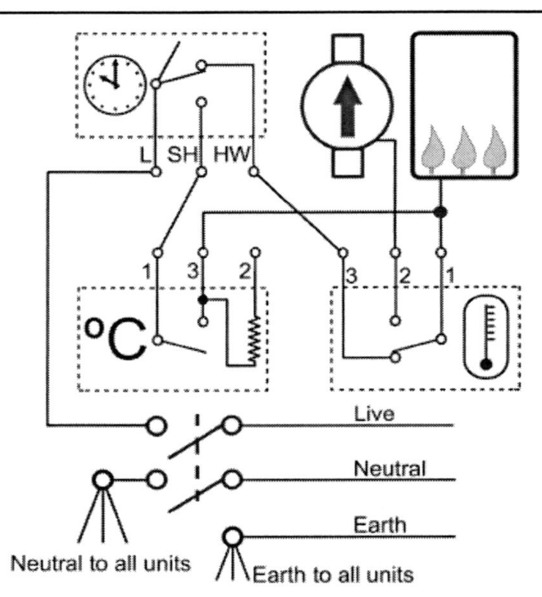

Figure 14.5: This is the best control circuit for partly pumped central heating systems using simple control components. It uses both a room thermostat and a tank thermostat and provides full boiler interlock.

How it works
When the mini-programmer is switch to hot water only the operation is the same as for the circuit of figure 14.4. If the tank thermostat is calling for heat it connects pin 1 to pin 3 and the boiler is on but the pump is not. When the mini-programmer is switched to both hot water and space heat both pin 1 of the room thermostat and pin 3 of the tank thermostat are live. If the tank thermostat only is calling for heat it connects pin 1 to pin 3 turning on the boiler but the pump is off. If the room thermostat is calling for heat and the tanks thermostat is not, the room thermostat connects pin 1 to pin 3 so the boiler is on and the tank thermostat connects pins 1 and 2 so the pump is on. If both thermostats are calling for heat the boiler is on but the pump is off because the tank thermostat is not connecting pin 2. This makes the system hot water priority. If neither thermostat is calling for heat both the pump and boiler are switched off giving boiler interlock.

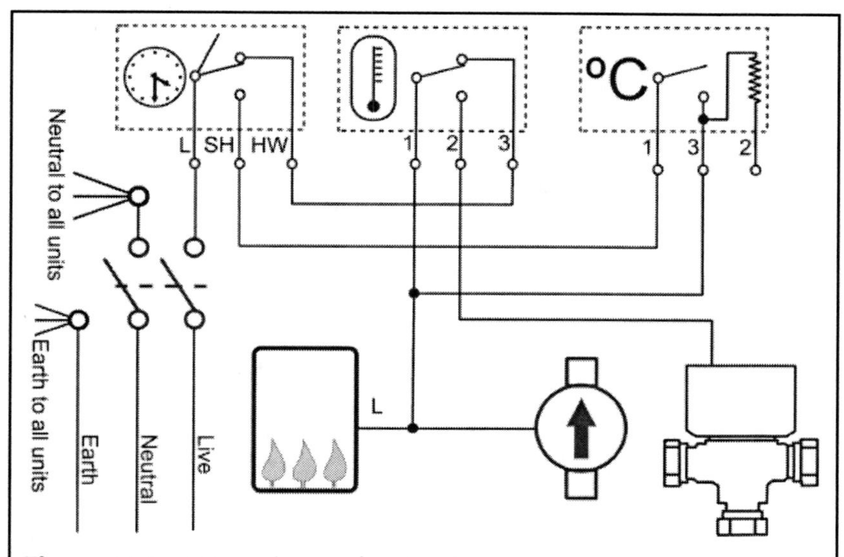

Figure 14.6: Wiring diagram for a W plan circuit

This is the control circuit for a fully pump system using a three port diverter valve and both room and tank thermostats. Unlike partly pumped systems, fully pump systems have the pump on whenever the boiler is on so they are both connected together.

How it works

The diverter valve when not energised diverts water to the primary hot water circuit and not the space heating. When the mini-programmer is switched to hot water, pin 3 of the tank thermostat is live. If the tank thermostat is calling for heat pin 3 is connected to pin 1 and both the boiler and pump are on. The diverter valve is not energised so is diverting water to the primary hot water circuit. If the tank thermostat is not calling for heat the boiler is off providing boiler interlock.

If the mini-programmer is switched to space heating and hot water pin 1 of the room thermostat is live. If it is calling for heat this is connected to pin 3 and both boiler and pump are on. Only when the tank thermostat is not calling for heat is pin 1 connected to pin 2, which energises the diverter valve moving it to the space heating position.

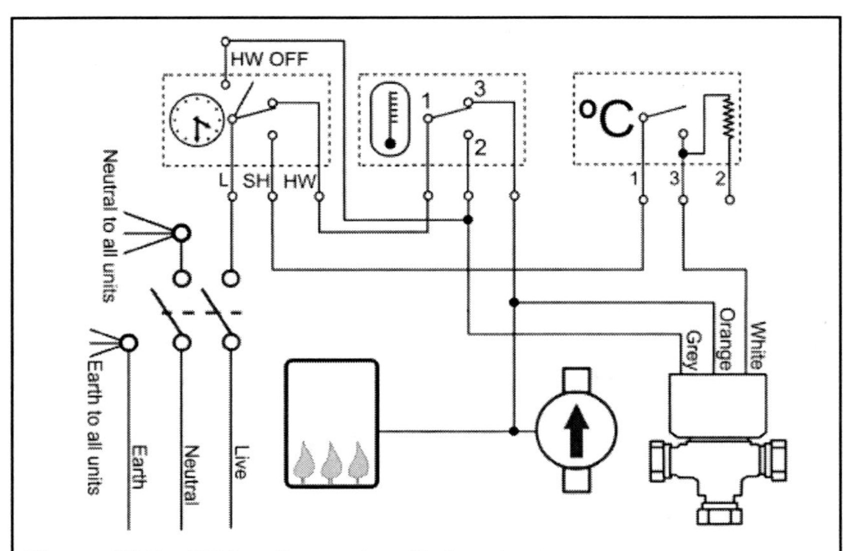

Figure 14.7: Wiring diagram for a Y plan circuit

The Y plan circuit uses a three port mid position valve which has ports A, B, and AB. When no power is connected to the valve port AB is connected with port B and this port is used for the hot water circuit. Port A is used for space heating. Unlike the previous circuits this circuit and those that follow will work properly with programmers that allow the space heating to be on without the hot water. Nowadays such programmers are referred to simply as programmers as opposed to mini-programs that do not allow this combination.

How it works:
There is an additional complexity used in this circuit. Programmers have a "hot water of" pin which is not used in other layouts. It goes live when the programmer is switched to hot water off. This is necessary for the operation of the mid position valve. When the programmer is in this position with hot water and space heating off, the grey input to the three port valve is live but the white is not. In this situation the valve retains whatever position it was in before the grey wire became live. This is likely to be with space heating off so the valve will be in its de-energised position.

When the programmer is switched to hot water on and space heating off, pin 1 of the tank thermostat will be live and the "hot water off" pin of the programmer will not be. If the tank thermostat is calling for heat pin 3 will be connected to pin 1. The orange wire to the three port valve is an output from it not an input so the valve will not be energised. The boiler and pump will be on and water will be diverted to the hot water circuit because the three port valve is in its de-energised position.

When the programmer is switched to hot water and space heating on, pin 1 of the tank thermostat and pin 1 of the room thermostat will be live. If the tank thermostat only is calling for heat its pin 3 will be live, the boiler and pump will be on, but the three port valve will not be energised. The water will be diverted round the hot water circuit. If the room thermostat only is calling for heat its pin 3 and the thank thermostat's pin 2 will be live. This means the grey and white inputs to the three port valve will be live. When both white and grey are live the valve moves to close port B and connect port AB to port A. When this happens the internal switch in the three port valve turns the orange wire to live (240V) turning on the boiler and pump. When both are calling for heat the grey input to the valve will not be live. Only the white will be live. In this condition the valve moves to its mid position. Water flows round both circuits.

When the programmer is switched to hot water off and central heating on the "hot water off" pin is live making the grey input of the valve live. If the room thermostat is calling for heat its pin 3 goes live making the white live and the valve moves fully to the A position connecting port AB to A only. When the room thermostat's call for heat is satisfied it turns off and the valve has a live on the grey and not on the white. In this situation it retains its previous position but the voltage on the orange wire falls from 240 to somewhere between 150 and 50. The system is designed so that this turns off the boiler and pump.

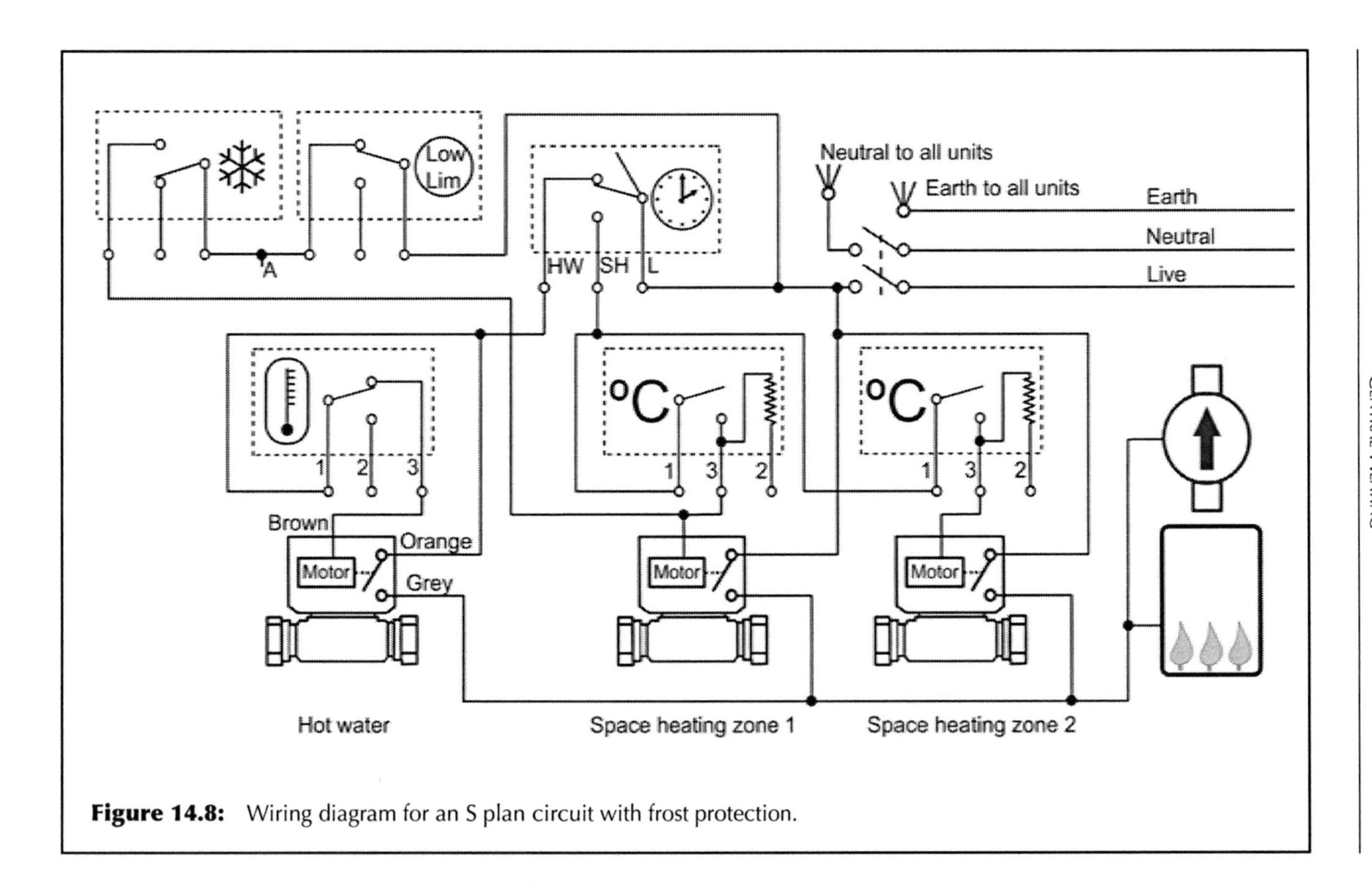

Figure 14.8: Wiring diagram for an S plan circuit with frost protection.

The S plan circuit uses 2 port valves. These incorporate a switch which is on when the valve is open.

How it works:

Ignore the low limit and frost thermostats for the moment. When the programmer is switched to hot water and space heating off, pin 1 of the tank thermostat and pin 1 of the room thermostats are not live. That means whatever their positions the inputs to the 2 port valve motors are not live. The valves are therefore closed and the valve switches off. Consequently the boiler and pump are also off.

When the programmer is switched to hot water on and space heating off (as shown) pin 1 of both room thermostats is not live so whatever their position the motors of the space heating 2 port valves will not be energised and their switches will be off. Pin 1 of the tank thermostat is live. If the tank thermostat is calling for heat then pin 1 is connected to pin 3. The hot water 2 port valve is activated and the valve opens and its switch turns on. This switch connects the input to the boiler and pump to the hot water output from the programmer which is live so the boiler and pump are on and water is diverted through the hot water circuit.

When the programmer is switched to both space heating and hot water on both outputs from the programmer are live and any thermostat calling for heat will activate its valve, turn on the valve's switch and turn on the boiler and pump. Notice that the boiler and pump will be on if one or more of the 2 port valve switches are on. If no thermostat is calling for heat than all valves are closed and all their switches are off so the boiler and pump will be off. This achieves boiler interlock.

Now look at the frost and low limit thermostats. The turn on temperature for the low limit thermostat is usually higher than that of the frost thermostat. As the temperature falls, first the low limit thermostat switches on. As the temperature falls further the frost thermostat switches on. Only when both thermostats are on is the frost protection system activated. Notice that the live input to the frost and low limit thermostats bypasses the programmer. Even if the programmer is switched off the input to the two port valve motor of space heating zone 1 will be activated. This valve will then open and the switch will be switched on. The input to this switch is not taken from the output from the programmer as is the case with the hot water system. Rather it is taken directly from the live immediately after the double pole switch bypassing the programmer. This means that even if the programmer is off when this valve switch turns on it will turn on the boiler and pump. Notice that only one of the space heating zones is activated directly by the frost protection system in this diagram. This will generally be the ground floor zone. Both zones will be activated if both room thermostats are calling for heat which is likely under these circumstances. There is no reason why further frost thermostat protection systems should not be added to other zones in which case the frost thermostats should be located in a place appropriate to their zone. For example if one zone is in an outbuilding then the frost thermostat for that zone should also be in the outbuilding. However many frost thermostats are used only one low limit thermostat is needed. The additional frost thermostats take their input from point "A".

Frost protection for other circuits
The S plan circuit of figure 14.8 has the advantage that the frost protection system fires the boiler and heats the relevant space heating circuit whether or not the room thermostats are calling for heat. Frost protection systems can be added to the other circuits above by wiring the low limit thermostat if used and frost thermostat together as shown in figure 15.8 but connecting pin 3 of the frost thermostat to the space heating output from the programmer. The frost system then works by bypassing the programmer only. The disadvantage of this is that the boiler will not be fired if the room thermostat is not calling for heat. In general under circumstances where a risk of frost damage exists the entire house will be very cold in which case thermostats would be calling for heat as their minimum setting when turned right down is 5 to 10°C but a risk still exists as they my not be placed near the coldest part of the system. A frost thermostat should be placed in an area at maximum risk of freezing rather than one representing a more average temperature. Near the lower most pipes against the most exposed wall is would be suitable.

These diagrams are of simple control systems using switches and electrical wiring only. More modern installations use sophisticated electronic programmers that bring a number of advantages in terms of reliability, efficiency, convenience, lack of radio interference, and flexibility. They perform the same basic functions as the switching circuits outlined above but their operation is not amenable to such ready analysis. Their installation is dependent on the detailed wiring instructions that come with the system. This can be a bit of a problem when maintaining or repairing systems but fortunately the Internet has brought ready access to detailed wiring diagrams of virtually every control system currently available or produced over the past few years.

CHAPTER 15
Other Heating Technologies

Solar heating

In hot climates solar heating is the standard way of providing hot water but in the British and Northern European climates it is not an economically competitive form of heating even with the help of the various grants available to promote the use of renewable energy. It has a niche in the market by virtue of its ecological advantages.

The basic problem is that there is not enough sunlight available in northern latitudes to provide anything more than moderate amounts of hot water during the summer. Specifically there is not enough to provide any degree of space heating during the winter when it is needed. It can only be used as a supplement to other heating systems but the modest saving in fuel bills it affords amount to less than the interest that the cost of its installation would generate in a deposit account!

Its market is largely restricted to people with a keen interest in the environment so solar heating equipment tends to be manufactured and installed in a very environmentally friendly way. Maximal use is made of recyclable materials. There is minimum waste of packaging. Installations often seek to make the maximum use of existing pipe work and tanks in order to avoid waste.

Solar heating systems are classified as low carbon or zero carbon. Low carbon systems use electric pumps and control systems powered from mains electricity. Zero carbon systems are entirely driven from solar power. A low carbon system will generally consume about 20% of the energy it absorbs from the sun to power pumps and control systems. This is usually taken from the mains electricity supply. Zero carbon systems either use photoelectric panels to drive an electric pump or convective currents to move water round the circuit. Convective circuits are very uncommon in the UK and Europe because they require that the hot water cylinder is mounted above the level of the solar panels. As the solar panels are usually mounted on the roof this is impractical. Zero carbon systems in Europe therefore usually use a photoelectric panel to generate electricity that drives a pump.

Heat pump systems

A heat pump is a device which moves heat from a cold place to a hot place. A fridge contains a heat pump that transfers heat from the inside of the fridge to the outside. Heat pumps used for heating remove heat from the outside environment and transfers it to the building. The source of heat is most commonly the ground but it can be atmospheric air or bodies of water.

The ground source heat pump system uses a series of pipes placed in the ground. A refrigeration plant cools these pipes, transferring the heat to the building. A certain amount of power is necessary to drive the compressors that operate the refrigeration plant but this is considerably less than the rate at which heat is transferred. Depending on the outside to inside temperature difference the system produces from 2 to 4 kW of useful heat per kW used to drive the compressors.

The big drawback of this system is its installation cost which at the time of writing would pay the gas bill for a gas fired heating system for between 30 and 60 years. The underground pipe work necessary is extensive. Pipes can either be buried in horizontal trenches or vertical bore holes. Boreholes range from 15 to 100 m deep or tens to hundreds of metres of trenches may be required. On the whole bore holes are preferable because the temperature of the ground is higher lower down and this improves the efficiency of the system.

Electric Systems

Heating buildings with standard mains electricity is expensive. Installation and maintenance costs are modest but in most situations energy costs outweigh the installation and maintenance savings that are made.

A feature of electrical generating technology makes electric heating more economic. It takes days to weeks to complete the shutdown or start-up of the large turbine driven generators that supply our electricity. Because of this considerable generating capacity has to be kept running 24 hours a day. This means that there is overcapacity during certain times such as the night when we are all asleep and not using much electricity. To encourage the use of this otherwise unused generating capacity the electricity companies offer cheap rate electricity for heating purposes during off peak hours. The exact hours vary but usually include the period between midnight and around 5 am and the early afternoon. The typical energy cost is a daily service charge that amounts to the price of about 2 standard kWh and about half the standard rate for off peak electricity used. At the time of writing this still makes it more expensive than gas or oil in the UK but by a smaller margin.

The higher energy costs of electric systems can be offset by other advantages which makes it economic in certain circumstances. They are highly reliable and require minimal maintenance. This can be a particular advantage in rented property where there is a legal requirement that gas fired systems are serviced annually. The servicing costs are usually borne by the landlord whereas the energy costs are borne by the tenant! Also electric systems are almost completely silent. There is more flexibility about where they can be installed because they have no requirement for a flue. They can be installed in areas where there is no mains gas supply but electricity demand is considerable and not all houses in rural areas have a sufficiently high-capacity electricity supply. Installation costs vary with systems but simple storage heater and hot water tank systems are cheap to install. The kind of circumstances where electricity is likely to be chosen are in holiday dwellings that are used mostly during summer months and have no mains gas supply and in rented accommodation particularly in small flats, and in buildings where gas is not available and accommodating an oil tank is problematic.

Wet electric heating

Electric heating systems fall into two groups: wet and dry. Wet systems are based on thermal storage tanks as described in chapter 3. In these systems a large volume of water is kept at low pressure in a very well insulated tank. This water is heated to a high temperature, typically 85°C. The heat is then removed from the water by a heat exchanger and given either to rising main water for hot taps or circulating water in a conventional space heating system. The water in the thermal store may be heated either by immersion heaters or by a primary hot water circuit heated by an electric boiler.

The heating power available is limited by the current capacity of the house. Most domestic houses have a capacity of around 15kW and electric boilers for domestic use are available in a range of powers up to this size. If higher powers are required then it's an easy matter to combine electric boilers in parallel.

Dry electric heating

Dry electric central heating systems use night storage heaters. These are heaters installed in each room against walls. They contain heating element surrounded by a number of high density high heat capacity bricks. These bricks make storage heaters extremely heavy. They are dangerous to handle without the appropriate equipment! The bricks are heated to a high temperature during the off peak period and this heat is then released at a variable rate by allowing convective air currents to flow over the bricks. The system is cheap and simple to install but suffers from being rather unresponsive. Also it is difficult to maintain an adequate heat output until the end of the day when it tends to be most needed. Storage heater systems are usually combined with direct hot water tanks. These are described in the chapter 3.

Dry electric heating is reliable and requires very little maintenance. It is inexpensive to install because there is no boiler or pipes. Of all heating systems it has the lowest installation costs and this fact has brought it a somewhat undeserved bad reputation. Unscrupulous landlords who need to install a heating system but will not be paying for the energy will install storage heaters. These same persons are likely to select the cheapest on the market. This has meant that many systems are poorly designed or matched to their houses with inadequate storage capacity making extensive use of top up heating with standard rate electricity necessary. This reputation is unfortunate because electric systems can be excellent if carefully chosen to have adequate capacity.

The design of both wet and dry systems involves a compromise between the amount of heat that can be stored versus space required and cost. This generally means that they store enough heat for daily use but not for periods of unusually high demand when top up heating will be required using standard rate electricity. The system is greatly enhanced if off peak electricity is available both during the night and the afternoon as then it does not need to store enough heat for the entire day. Further details of thermal storage systems are given in chapter 3.

The use of off peak electricity requires that the electricity company installs a special kind of meter in the house which adjusts the rate charged according to the time of day. In the past, off peak storage systems used two separate circuits. The storage heaters and hot water were connected to one with a time switch and the rest of the electricity supply to the house was connected to another. This increased the installation cost as additional wiring had to be run all round the house. Modern systems use one circuit but adjust the tariff according to the time. Appliances that use off peak electricity are equipped with timing devices that switch them on at the appropriate times of day.

Electric showers

Showers are a very popular form of electric heating because they are inexpensive to install, provide immediate hot water with no need to wait for a tank to heat up and allow contingency against failure of the hot water system.

They are very powerful electric devices and cannot be supplied by the common 2.5mm^2 twin and earth cable that carries current to mains sockets. Six mm^2 cable as used for cookers is required. People using a shower are wet and unclothed making them particularly vulnerable to electric shocks so extra safety precautions are necessary. In the UK newly installed electric showers must be connected via an RCD protector. Many consumer units in modern houses have RCD protection and where they haven't an RCD protected shower consumer unit must be installed with a new shower. In addition to this is the shower must also be controlled by a double pole switch the positioning of which is subject to various regulations (figure 15.1).

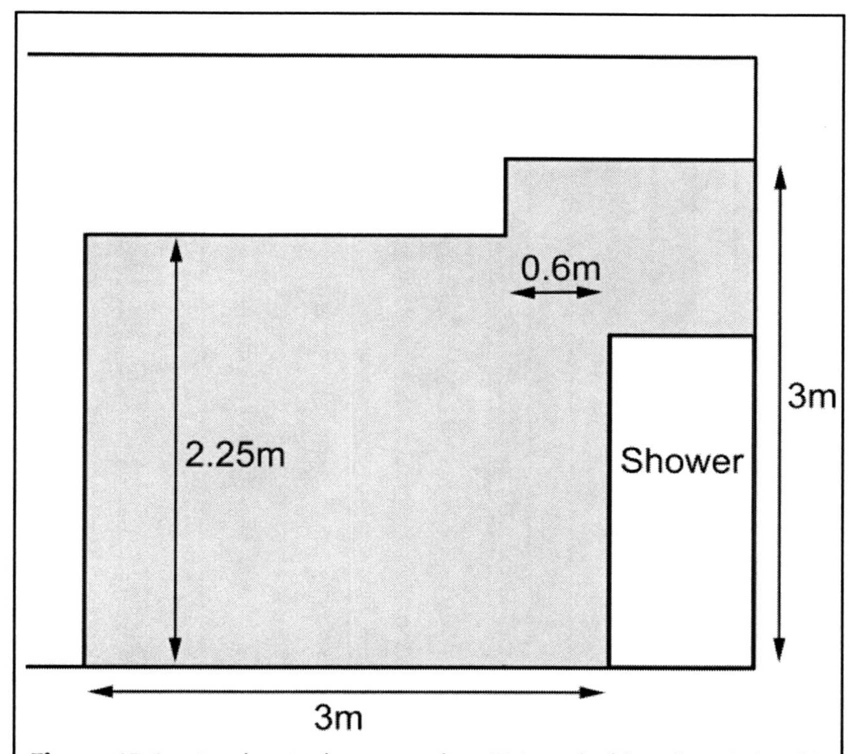

Figure 15.1: An electric shower needs a 45 Amp double pole switch. This should be located either more than 3m above the floor, or more than 2.25m above the floor and more than 0.6 m horizontally from the shower or more than 3m horizontally from the shower; i.e. not in the grey area. A pull switch can have an insulating pull cord (not metal chain!) that hangs into the grey area but not into the shower itself.

CHAPTER 16
Fault Diagnosis

Central heating fault diagnosis has undergone a major change over the last two decades because of the introduction of computerised control and on board diagnostic functions. This trend has turned the boiler into a "black box" whose malfunctions can only be understood with dedicated diagnostic equipment or a list of specific manufacturer's codes. The range of boilers encountered spans the period over which these changes took place and specialist diagnostic skills are still necessary but increasingly the diagnostic process is restricted to identifying and solving faults which occur out-with the boiler and decoding and correcting faults that are identified by on board diagnostic software within the boiler casing.

Many faults do not arise because of a manufacturing defect or because a part has reached the end of its working life but because of another fault in the system. It is very frequent for parts to be replace under these circumstances only to fail shortly afterwards if the underlying fault remains. I recommend that you critically asked the question what caused the component to fail. If the answer is obvious, for example a pump has failed at the end of its design life, it is not necessary to be suspicious of some provoking factor but regard this as the exception rather than the rule.

Boiler faults

If the input of the boiler is live but the boiler won't fire it is at fault. There is quite often some kind of indicator light on the boiler which enables you to tell whether the supply of electric current is on or not without having to test for it. Fuses in central heating systems frequently blow because of the prolonged periods they are on and because when the pump motor starts up it briefly draws a high current. Check that the gas is switched on. The handle of the main gas tap should be pointing in the direction of the gas flow rather than at 90° to it. Sometimes after work has been done on the gas supply the pipe may be full of air having not yet been purged.

Pilot lights

Gas burning appliances are designed to carefully mix quantities of gas and air to optimise the efficiency of combustion. If for some reason the burner should not light or go out but the gas and air mixture continues to flow it will fill the space around the burner and quite possibly the room. In that case a bomb exists waiting for a spark, such as from a light switch, to ignite a large volume of perfectly balanced gas and air leading to an explosion. To prevent this disaster the system has to ensure that the burner is lit whenever gas is flowing and this is done with a pilot light.

The operation of a pilot light is shown in figure 16.1. The mains gas piped to both the pilot light and the main burner is controlled by a valve. This valve opens the flow of gas only if a temperature sensor called a flame failure device in the pilot light flame is hot. In order to start the pilot light an auxiliary pathway for gas flow is needed. This is either controlled manually or automatically using a strictly time-limited switch. In the manual system the user presses a button which has the dual functions opening the auxiliary gas pathway and firing a spark that ignites the pilot light. The user then has to keep pressing the button until the flame failure device has heated up and opened the main gas valve which usually takes a few seconds. Only then can the button be released as the gas will continue to flow through the main valve. This should only be necessary after boiler shutdown as the pilot light is always lit in normal operation. The pilot light is positioned so that it must ignite gas flowing from the main burner. If the pilot light goes out the flame failure device in its flame cools and turns off the main gas valve. Thus if the pilot light is lit the burner will light, if the pilot light is out the valve will not open so unlit gas flow is prevented.

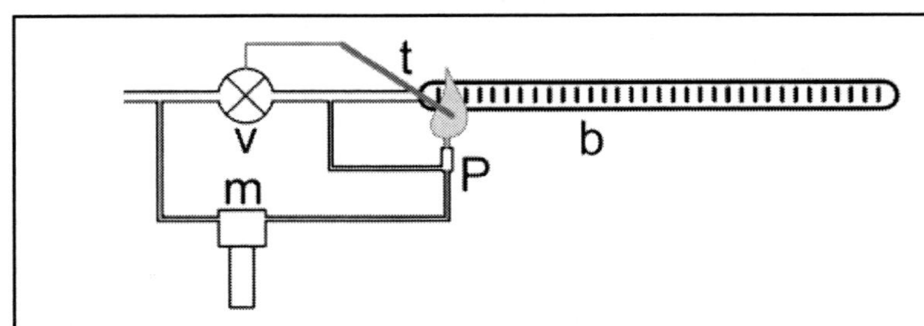

Figure 16.1: The pilot light safety system. The pilot light (p) is directed across the main burner (b) in order to always ignite gas coming from it. A temperature sensor (t) known as the flame failure device is in the pilot light flame and controls the main burner gas valve (v). The valve will only open when the sensor is hot. To light the pilot light a manual valve (m) that allows slow gas flow only when the button is pressed in is used.

There are several common failure modes of the system which result in the boiler not firing. Together they account for the large majority of boiler breakdowns. The pilot light flame may not be directed accurately towards the flame failure device. This can happen if the pilot light jet is partially blocked diverting the flame to the side or if the flame failure device is bent out of the way the flame. The solution is to clear the jet or bend the sensor back into the flame. The flame failure device itself has a limited life and may fail. For reasons of safety it is designed to fail giving a cold signal. It is a readily available replacement part and can be fitted by any competent person.

Another related cause of failure is of the piezoelectric ignition spark generator in manual systems which is quite common.

Modern boilers have a number of safety control systems to prevent the main burner from firing if it would lead to danger or damage. These use a variety of sensors to detect things like no water in the heat exchanger or a blocked flue. The systems are designed to fail safe so the boiler will not fire if either a problem is detected or a sensor is malfunctioning. As the result is the same for all such failures, you have to rely on on-board diagnostics to find the fault.

Pump failure

Pumps fail in a limited number of ways. They can seize, burnout, or the bearings can wear out. The symptom of bearing wear is primarily noise rather than failure but ultimately it will lead to seizure. If the pump seizes but the electric supply continues eventually the insulation of the windings will burn out. All of these modes of failure require that the pump is replaced but this is not always needed. If the pump has been inactive for a long time, for example over the summer when heating is not in use, the bearings may stick such that the pump does not start but can be easily freed with a screwdriver as described in chapter 10.

Control system faults

The traditional mains voltage control systems described in chapter 12 are slowly giving way to electronic systems where thermostats are low voltage devices that are either connected directly to the controller or battery operated with no physical connection to the programmer, sending temperature information by a radio link. Such systems tend to be reliable. When they do fail diagnosis cannot usually be taken any further than identifying which component has failed and in cases of doubt this may need to be done by swapping components for a known good one to see if the fault is fixed. There are two “faults” that you should bear in mind. A flat battery in a component such as a thermostat is a common

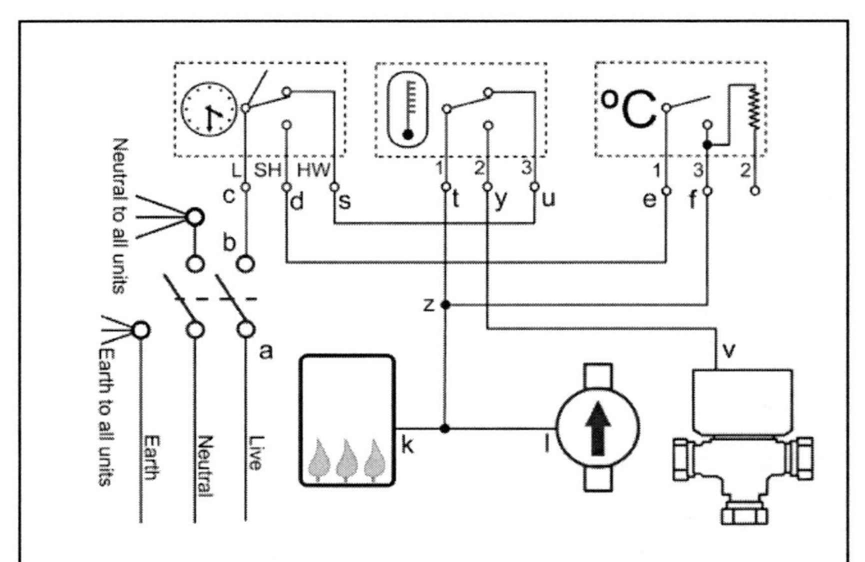

Figure 16.2: The diagram of figure 14.6 reproduced with letter labelling of points in the circuit for fault tracing as per the flow diagrams.

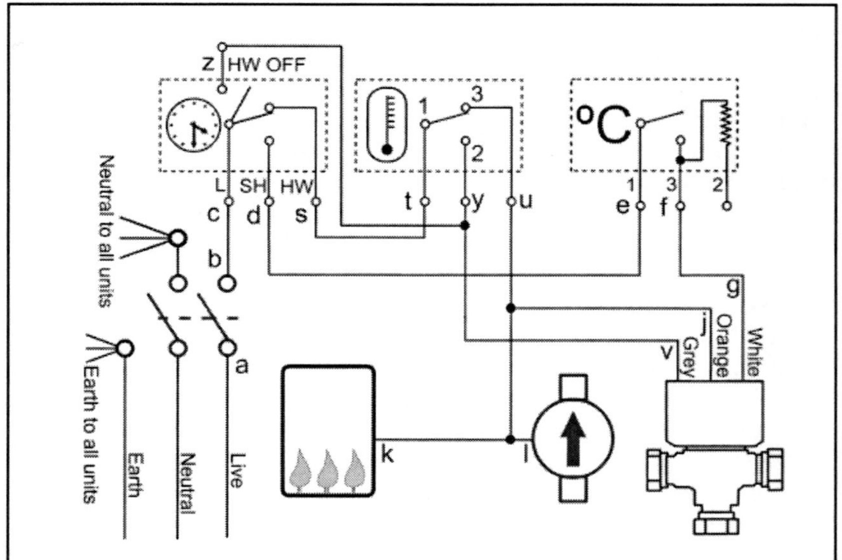

Figure 16.3: The diagram of figure 14.7 reproduced with letter labelling of points in the circuit for fault tracing as per the flow diagrams.

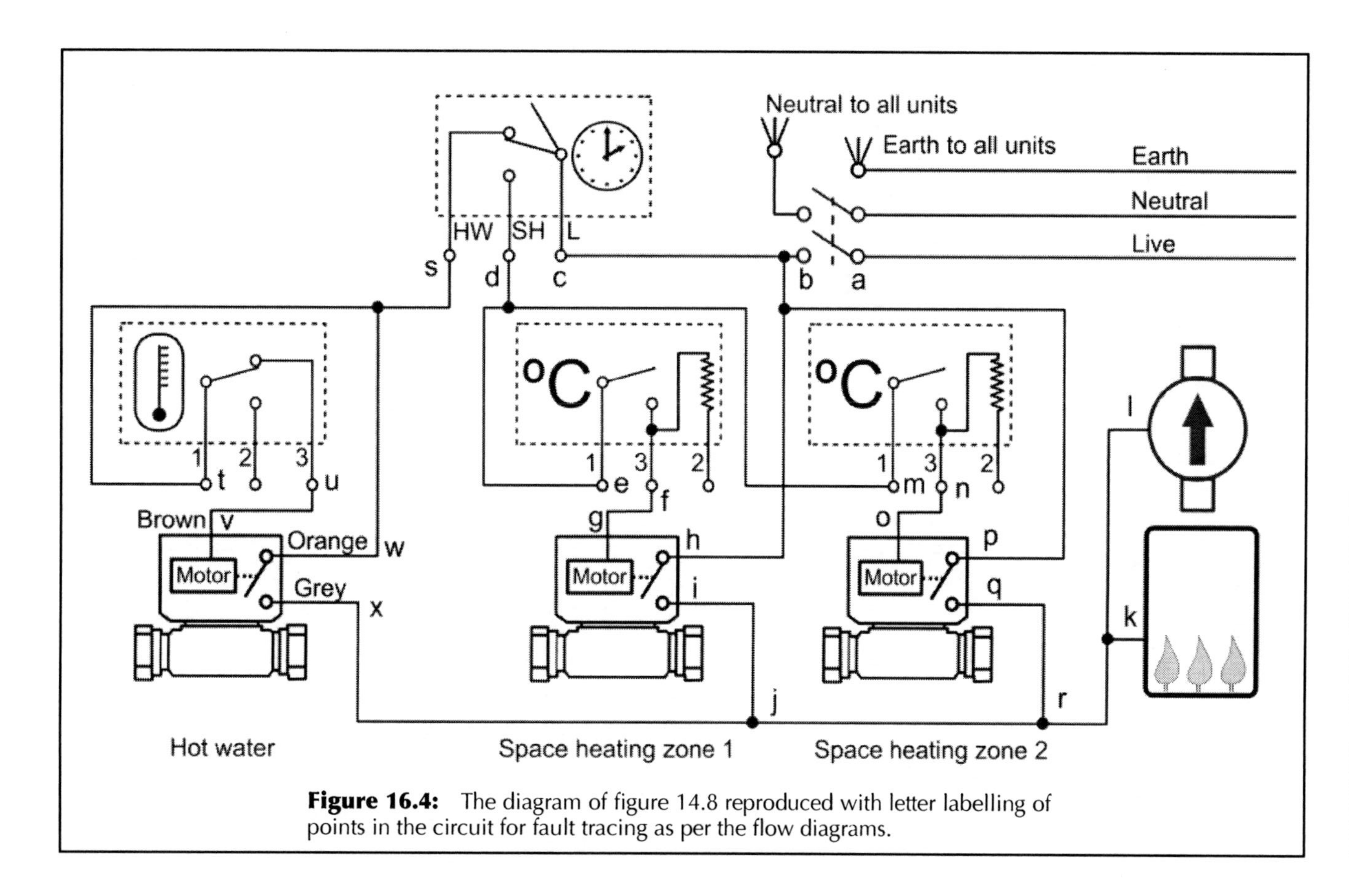

Figure 16.4: The diagram of figure 14.8 reproduced with letter labelling of points in the circuit for fault tracing as per the flow diagrams.

problem especially as users often do not realise that batteries are needed. The second is inappropriate programming. Electronic programmers have numerous sophisticated features often controlled by relatively few buttons making the process of programming them complex. Reprogramming can be achieved by an inquisitive child though the result may not be optimal system operation. When this happens it may be quite beyond the users to fix, especially without the control system's documentation.

Fault diagnosis in mains voltage systems is a different problem entirely. For one thing the combination of mains electricity, copper piping and water makes the potential hazards of handling the control system very great. Special precautions are necessary and it is recommended an RCD protector is used in the circuit supplying the system. The other problem is that mains systems are complex to assemble. Not only are all the parts prone to failure but incorrect wiring and assembly leading to malfunctions is common. As in chapter 13 most of the discussion on control systems in this chapter focuses on mains voltage systems.

When electrical faults occur it is almost always on the live side of the circuit. The neutral is rarely a problem because it is not connected through a series of failure prone devices. It can be checked by measuring the voltage between any known good live and the neutral in question which should read 240VAC.

In general the live side is inspected by tracing the current path from the input to the boiler back through the various thermostats and switches ultimately to the double pole switch controlling the mains supply to the central heating system. Each available point is tested to see if it is live or not. If the problem is that the boiler and pump won't turn on then at some point round the circuit the reading will change from not live to live. The wiring link or unit between the not live and live points is at fault. As an example supposing we have an S plan circuit as shown in figure 16.3. The live side is tested by checking the voltage throughout the circuit from points k and l to point a where a is the input to the double pole switch, b is the output from the double pole switch, c is the input to the programmer etc. They do not have to be tested in sequence. One way of doing it would be to divide the circuit up into bits and test perhaps i first. If this is live the fault must lie between j and l. If it is not live and both room thermostats are calling the heat the fault must lie between a and i or between p and r. Another approach would be to start with all the points that are close together. Quite frequently the pump, boiler and motorised valves can be conveniently tested at the same time.

A complication that arises with some circuits is that a point may be live for more than one reason. For example in figure 16.3 of the Y plan circuit point v may be live because the programmer is switched to "hot water off" or because the tank thermostat is not calling for heat and so switched to connect its pins 1 and 2. If v should not be live either could have failed. It may be necessary to disconnect them in turn to trace the fault correctly.

Below are a series of diagnostic flow diagrams aimed at mains voltage systems. They are each for specific symptoms and give a guide to the diagnostic approach to take. They are not absolute as some variation occurs between systems. The key tool in diagnosis is a circuit diagram and a logical approach to narrowing down the fault as far as possible without voltage testing first, then doing the minimum voltage testing or disconnections necessary to pinpoint the fault.

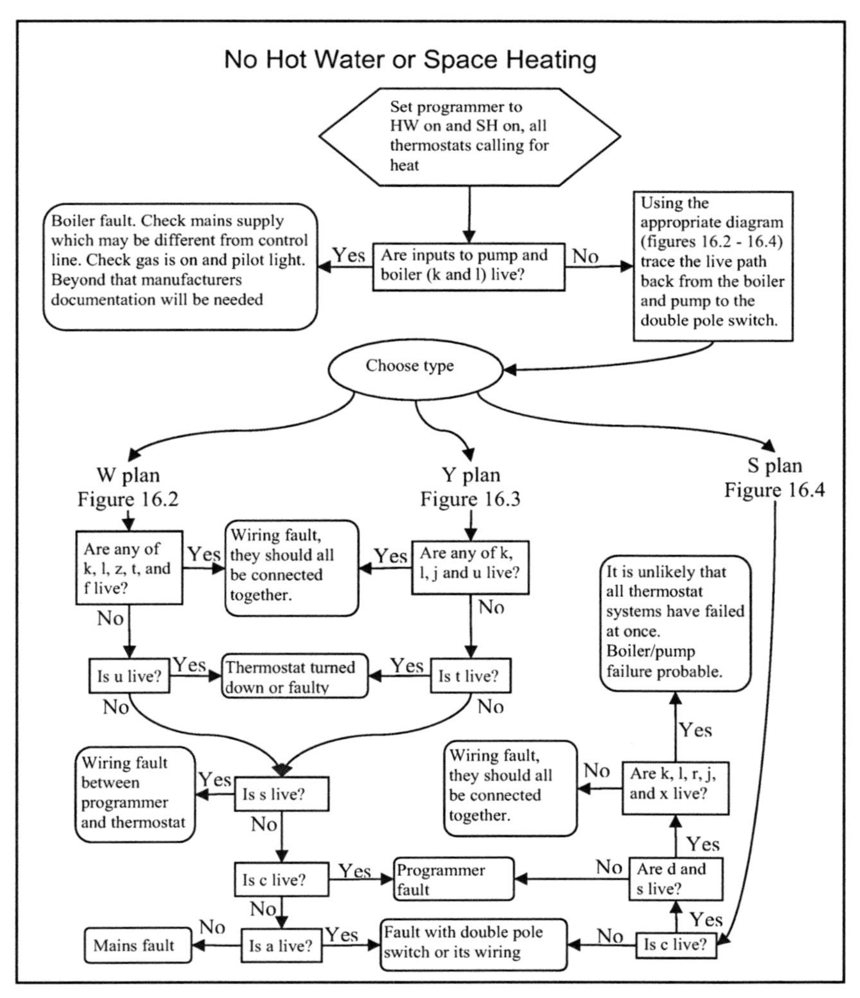

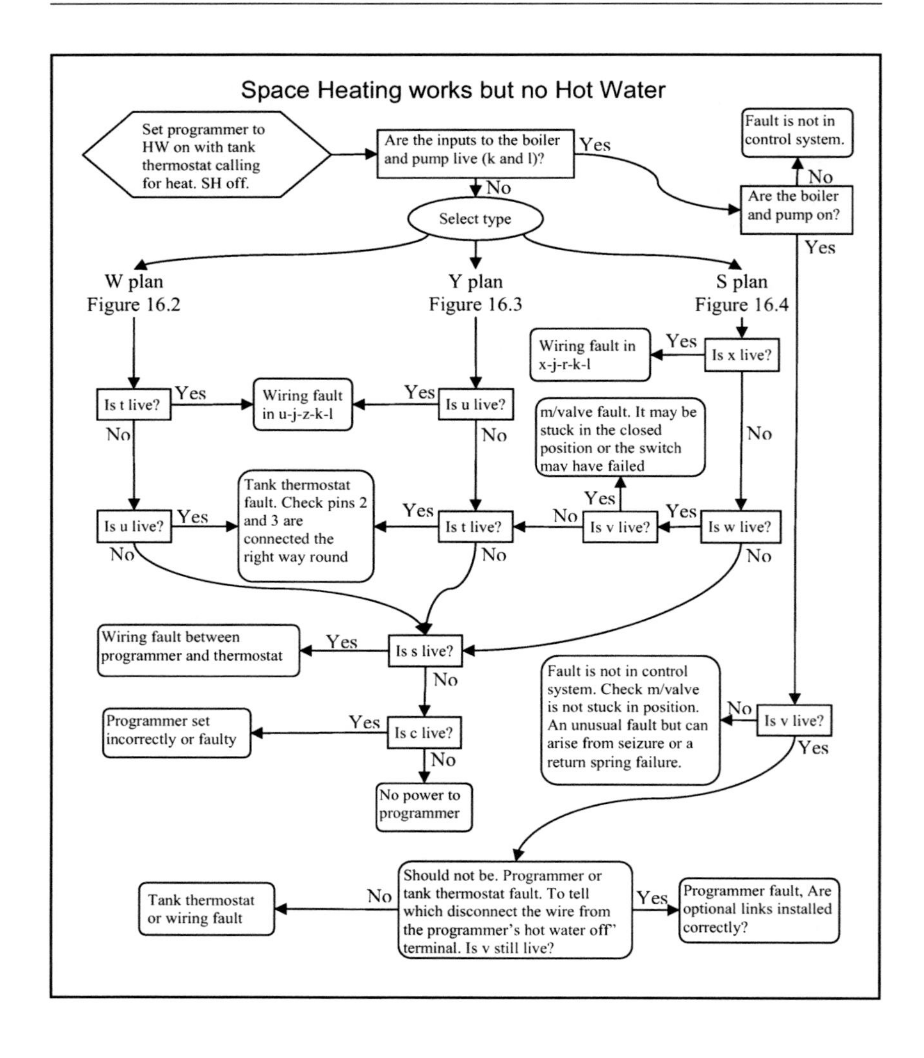
Space Heating works but no Hot Water
Set programmer to HW on with tank thermostat calling for heat. SH off.
Are the inputs to the boiler and pump live (k and l)?
Yes
No
Fault is not in control system.
No
Are the boiler and pump on?
Yes
Select type
W plan Figure 16.2
Y plan Figure 16.3
S plan Figure 16.4
Wiring fault in x-j-r-k-l
Yes
Is x live?
Is t live?
Yes
Wiring fault in u-j-z-k-l
Yes
Is u live?
No
No
m/valve fault. It may be stuck in the closed position or the switch may have failed
No
Tank thermostat fault. Check pins 2 and 3 are connected the right way round
Yes
Is u live?
Yes
Is t live?
No
Is v live?
Yes
Yes
Is w live?
No
No
No
Wiring fault between programmer and thermostat
Yes
Is s live?
No
Fault is not in control system. Check m/valve is not stuck in position. An unusual fault but can arise from seizure or a return spring failure.
No
Is v live?
Yes
Programmer set incorrectly or faulty
Yes
Is c live?
No
No power to programmer
Tank thermostat or wiring fault
No
Should not be. Programmer or tank thermostat fault. To tell which disconnect the wire from the programmer's hot water off' terminal. Is v still live?
Yes
Programmer fault, Are optional links installed correctly?

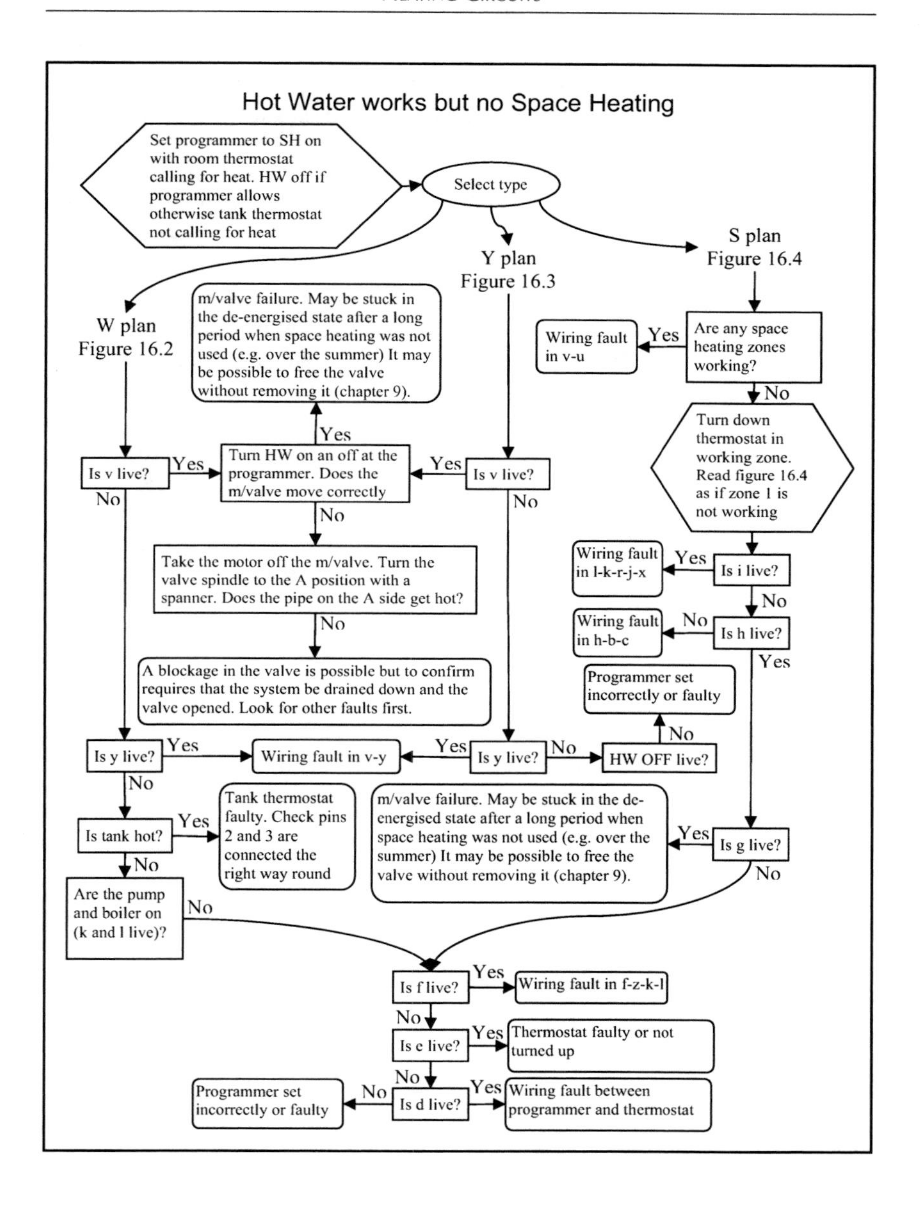
Hot Water works but no Space Heating
Set programmer to SH on with room thermostat calling for heat. HW off if programmer allows otherwise tank thermostat not calling for heat
Select type
W plan Figure 16.2
Y plan Figure 16.3
S plan Figure 16.4
m/valve failure. May be stuck in the de-energised state after a long period when space heating was not used (e.g. over the summer) It may be possible to free the valve without removing it (chapter 9).
Wiring fault in v-u
Yes
Are any space heating zones working?
No
Turn down thermostat in working zone. Read figure 16.4 as if zone 1 is not working
Is v live?
Yes
Turn HW on an off at the programmer. Does the m/valve move correctly
Yes
Is v live?
No
No
No
Take the motor off the m/valve. Turn the valve spindle to the A position with a spanner. Does the pipe on the A side get hot?
No
Wiring fault in l-k-r-j-x
Yes
Is i live?
No
Wiring fault in h-b-c
No
Is h live?
Yes
A blockage in the valve is possible but to confirm requires that the system be drained down and the valve opened. Look for other faults first.
Programmer set incorrectly or faulty
No
Is y live?
Yes
Wiring fault in v-y
Yes
Is y live?
No
HW OFF live?
No
Is tank hot?
Yes
Tank thermostat faulty. Check pins 2 and 3 are connected the right way round
m/valve failure. May be stuck in the de-energised state after a long period when space heating was not used (e.g. over the summer) It may be possible to free the valve without removing it (chapter 9).
Yes
Is g live?
No
No
Are the pump and boiler on (k and l live)?
No
Is f live?
Yes
Wiring fault in f-z-k-l
No
Is e live?
Yes
Thermostat faulty or not turned up
No
Programmer set incorrectly or faulty
No
Is d live?
Yes
Wiring fault between programmer and thermostat

Hot Water or Space Heating too hot

These faults usually arise for one of two reasons: a thermostat fault or a seized motorised valve. Thermostat faults are those where the thermostat continues to call for heat when no more is required. This may be because the thermostat is cooler than the room or hot water it is meant to control. A poor thermal contact, or no contact, between a tank thermostat and the hot water cylinder is a common cause of water being to hot. The thermostat switch may be stuck in one position or there may be a wiring fault.

Motorised valve faults cause too much heat when one thermostat (let us say the room thermostat) is calling for heat and another (let us say the tank thermostat) is not. The boiler is on to supply space heating but the water is being diverted to the wrong place by a defective valve. Consequently this fault is usually associated too much heat in one system and not enough in another. The commonest problem is that the hot water is too hot and there is no space heating. This is because when motorised valves fail they usually do so in the hot water (de-energised) position. The flow diagrams below outline the diagnostic approach.

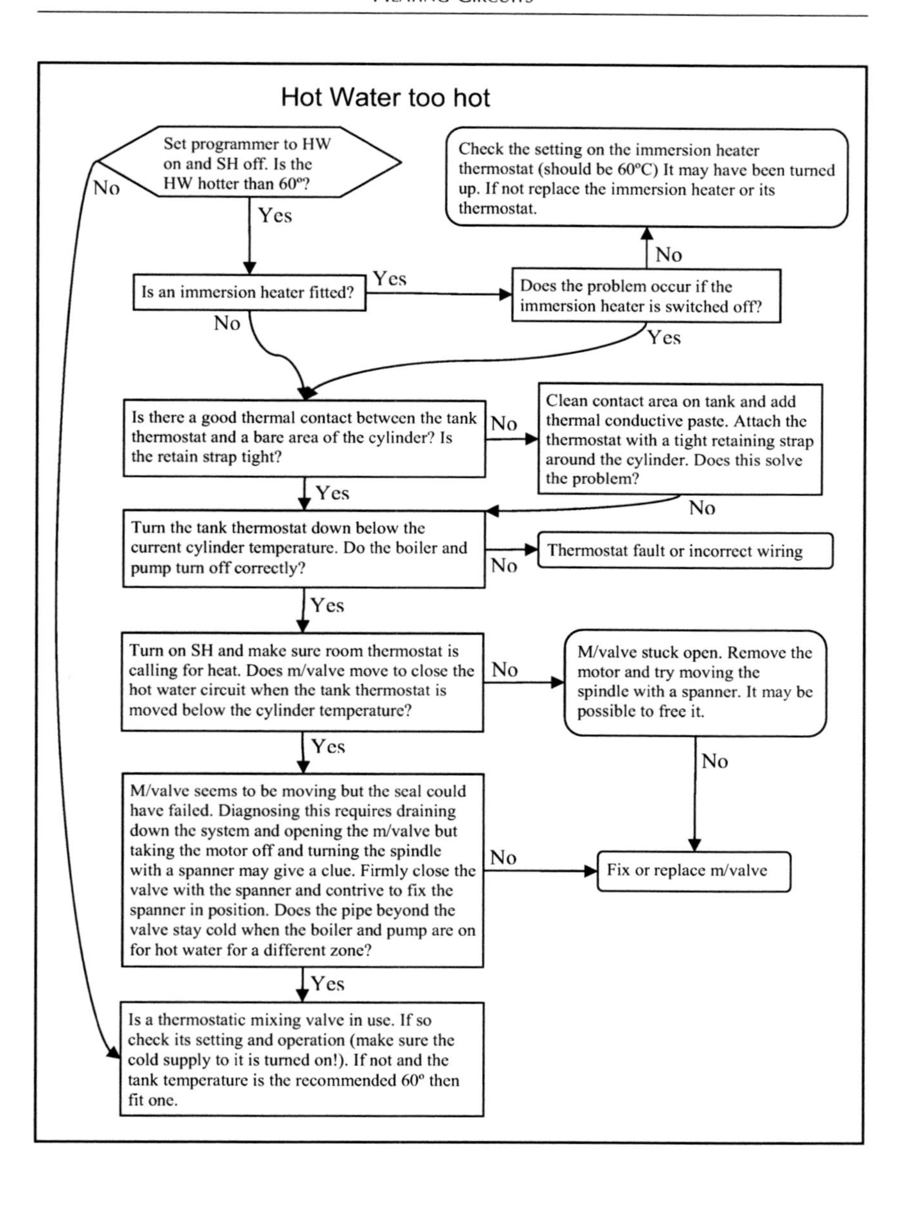
Hot Water too hot
Set programmer to HW on and SH off. Is the HW hotter than 60°?
No
Yes
Check the setting on the immersion heater thermostat (should be 60°C) It may have been turned up. If not replace the immersion heater or its thermostat.
No
Is an immersion heater fitted?
Yes
Does the problem occur if the immersion heater is switched off?
No
Yes
Is there a good thermal contact between the tank thermostat and a bare area of the cylinder? Is the retain strap tight?
No
Clean contact area on tank and add thermal conductive paste. Attach the thermostat with a tight retaining strap around the cylinder. Does this solve the problem?
Yes
No
Turn the tank thermostat down below the current cylinder temperature. Do the boiler and pump turn off correctly?
Thermostat fault or incorrect wiring
No
Yes
Turn on SH and make sure room thermostat is calling for heat. Does m/valve move to close the hot water circuit when the tank thermostat is moved below the cylinder temperature?
No
M/valve stuck open. Remove the motor and try moving the spindle with a spanner. It may be possible to free it.
Yes
No
M/valve seems to be moving but the seal could have failed. Diagnosing this requires draining down the system and opening the m/valve but taking the motor off and turning the spindle with a spanner may give a clue. Firmly close the valve with the spanner and contrive to fix the spanner in position. Does the pipe beyond the valve stay cold when the boiler and pump are on for hot water for a different zone?
No
Fix or replace m/valve
Yes
Is a thermostatic mixing valve in use. If so check its setting and operation (make sure the cold supply to it is turned on!). If not and the tank temperature is the recommended 60° then fit one.

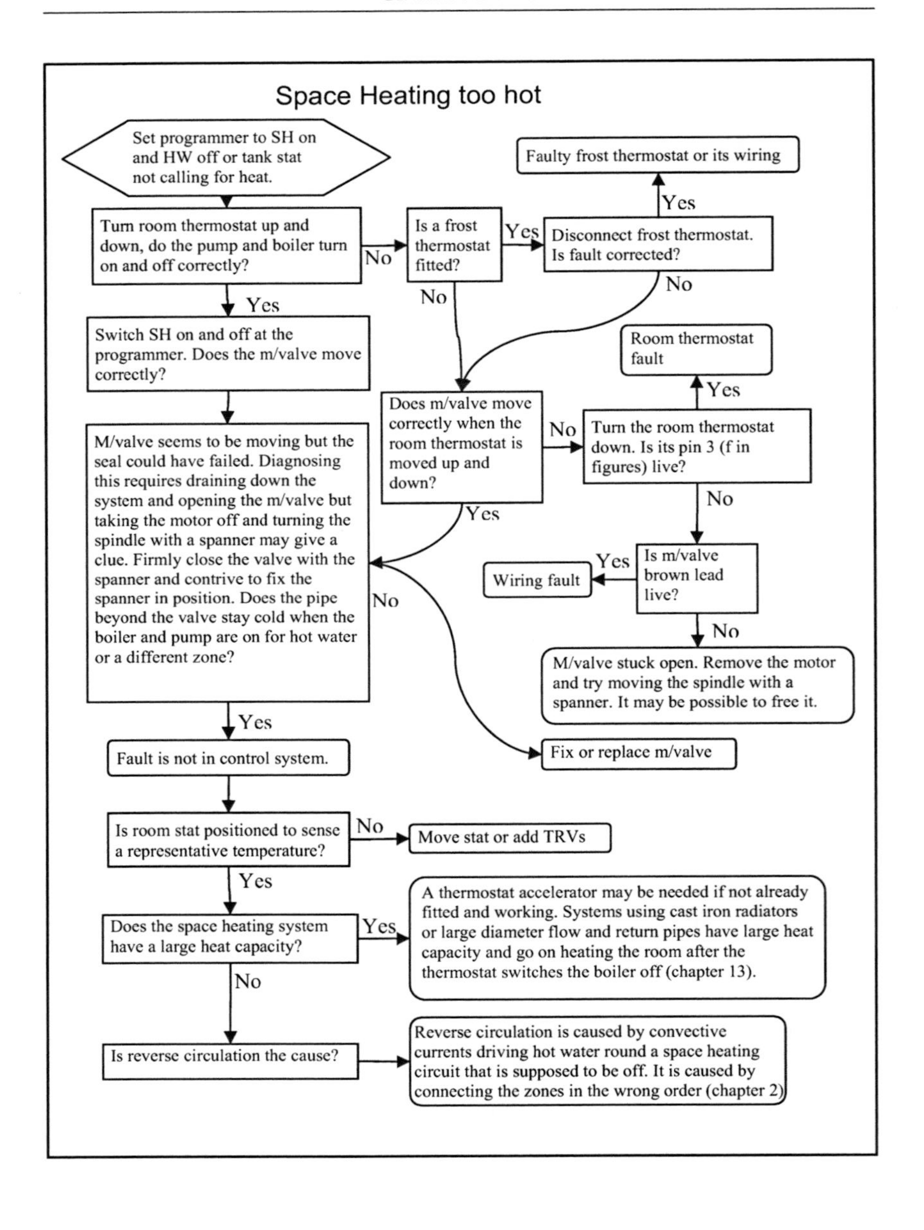
Space Heating too hot
Set programmer to SH on and HW off or tank stat not calling for heat.
Turn room thermostat up and down, do the pump and boiler turn on and off correctly?
No
Is a frost thermostat fitted?
Yes
Disconnect frost thermostat. Is fault corrected?
Yes
Faulty frost thermostat or its wiring
No
No
Yes
Switch SH on and off at the programmer. Does the m/valve move correctly?
M/valve seems to be moving but the seal could have failed. Diagnosing this requires draining down the system and opening the m/valve but taking the motor off and turning the spindle with a spanner may give a clue. Firmly close the valve with the spanner and contrive to fix the spanner in position. Does the pipe beyond the valve stay cold when the boiler and pump are on for hot water or a different zone?
Does m/valve move correctly when the room thermostat is moved up and down?
No
Turn the room thermostat down. Is its pin 3 (f in figures) live?
Yes
Room thermostat fault
No
Is m/valve brown lead live?
Yes
Wiring fault
No
M/valve stuck open. Remove the motor and try moving the spindle with a spanner. It may be possible to free it.
Yes
No
Fix or replace m/valve
Yes
Fault is not in control system.
Is room stat positioned to sense a representative temperature?
No
Move stat or add TRVs
Yes
Does the space heating system have a large heat capacity?
Yes
A thermostat accelerator may be needed if not already fitted and working. Systems using cast iron radiators or large diameter flow and return pipes have large heat capacity and go on heating the room after the thermostat switches the boiler off (chapter 13).
No
Is reverse circulation the cause?
Reverse circulation is caused by convective currents driving hot water round a space heating circuit that is supposed to be off. It is caused by connecting the zones in the wrong order (chapter 2)

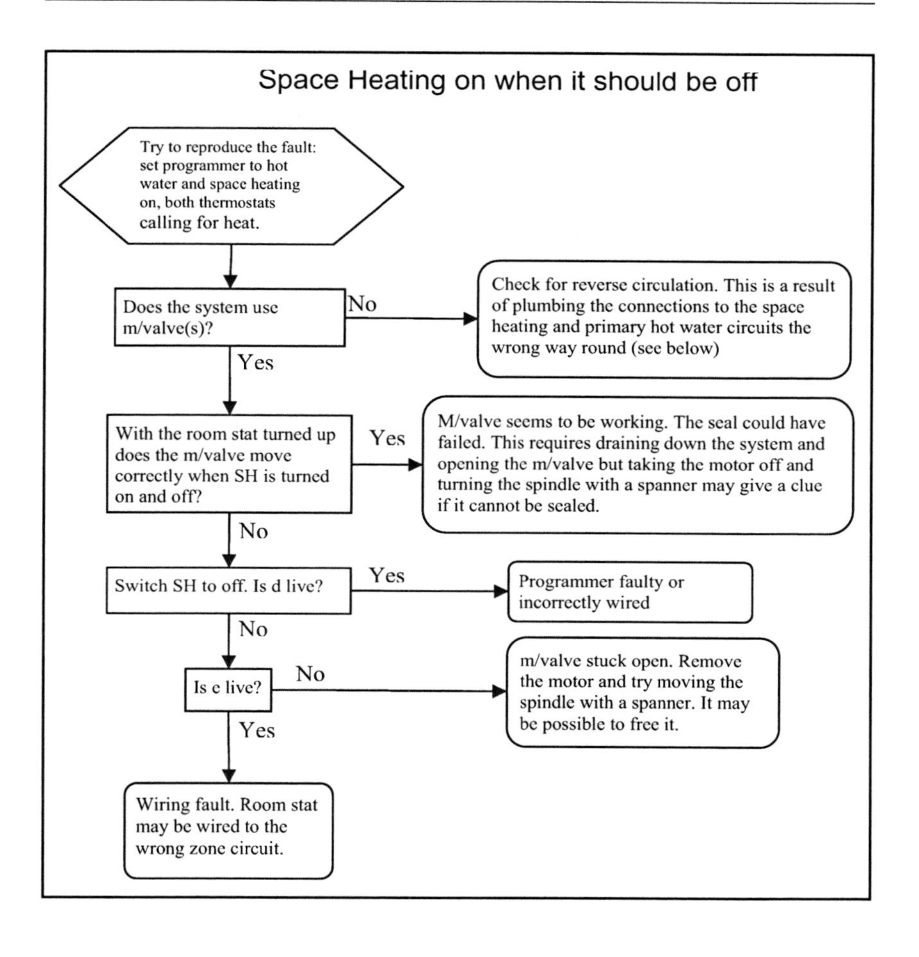
Space Heating on when it should be off
Try to reproduce the fault: set programmer to hot water and space heating on, both thermostats calling for heat.
Does the system use m/valve(s)?
No
Check for reverse circulation. This is a result of plumbing the connections to the space heating and primary hot water circuits the wrong way round (see below)
Yes
With the room stat turned up does the m/valve move correctly when SH is turned on and off?
Yes
M/valve seems to be working. The seal could have failed. This requires draining down the system and opening the m/valve but taking the motor off and turning the spindle with a spanner may give a clue if it cannot be sealed.
No
Switch SH to off. Is d live?
Yes
Programmer faulty or incorrectly wired
No
Is e live?
No
m/valve stuck open. Remove the motor and try moving the spindle with a spanner. It may be possible to free it.
Yes
Wiring fault. Room stat may be wired to the wrong zone circuit.

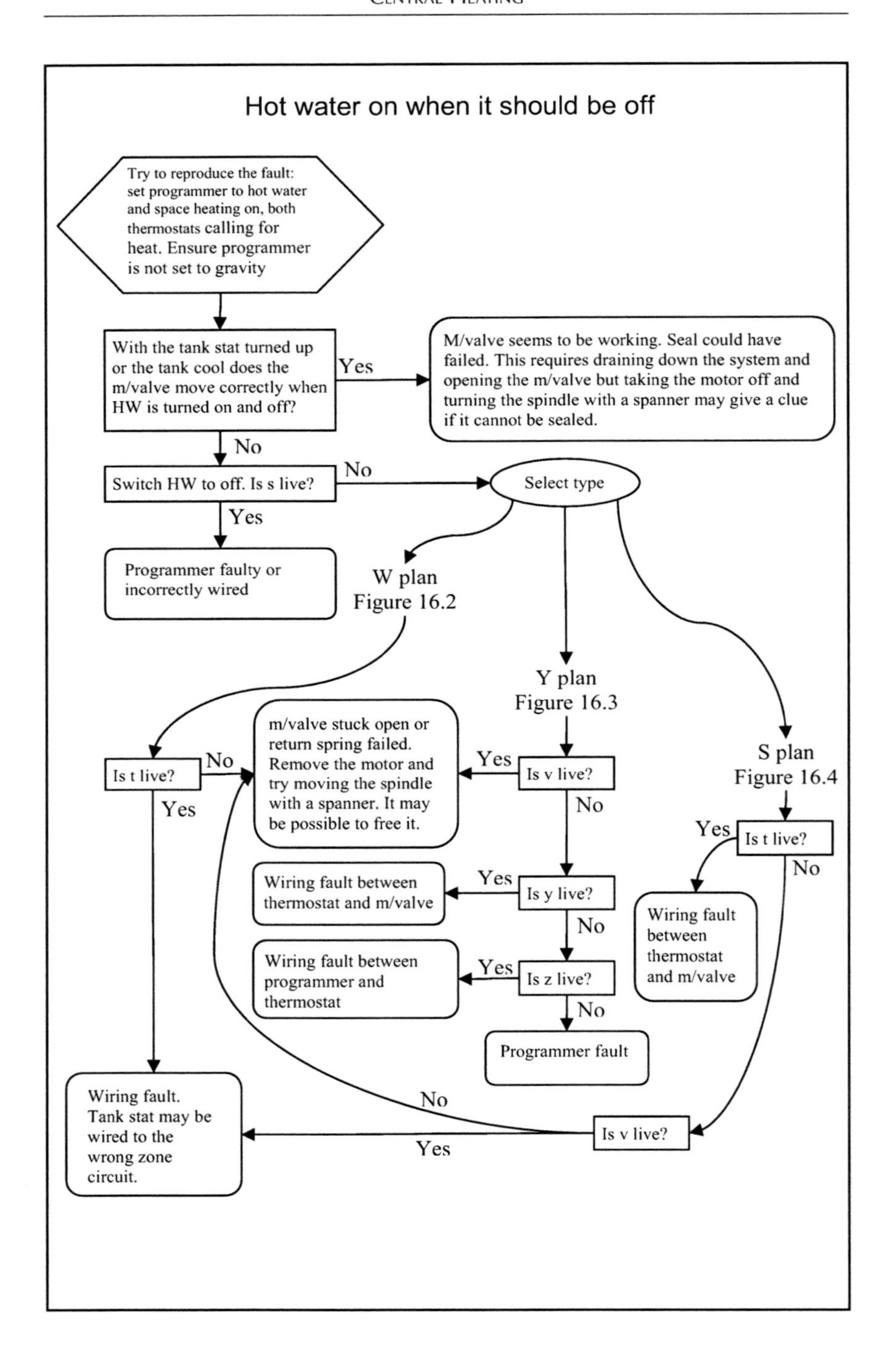
Hot water on when it should be off
Try to reproduce the fault: set programmer to hot water and space heating on, both thermostats calling for heat. Ensure programmer is not set to gravity
With the tank stat turned up or the tank cool does the m/valve move correctly when HW is turned on and off?
Yes
M/valve seems to be working. Seal could have failed. This requires draining down the system and opening the m/valve but taking the motor off and turning the spindle with a spanner may give a clue if it cannot be sealed.
No
Switch HW to off. Is s live?
No
Select type
Yes
Programmer faulty or incorrectly wired
W plan
Figure 16.2
Y plan
Figure 16.3
S plan
Figure 16.4
Is t live?
No
Yes
m/valve stuck open or return spring failed. Remove the motor and try moving the spindle with a spanner. It may be possible to free it.
Yes
Is v live?
No
Yes
Is t live?
No
Wiring fault between thermostat and m/valve
Yes
Is y live?
No
Wiring fault between thermostat and m/valve
Wiring fault between programmer and thermostat
Yes
Is z live?
No
Programmer fault
Wiring fault. Tank stat may be wired to the wrong zone circuit.
No
Is v live?
Yes

Noise

Most of the time when a central heating system goes wrong the problem is not difficult to trace but for some reason this is not true of noise. Frequently a central heating systems making an abnormal noise is subjected to numerous replacement parts and fiddling without any solution being found. This is the more surprising as most causes of noise can be recognised from the type of sound they make. The difficulty is usually not that the diagnosis is hard to make, but that the noise sounds as though it is coming from one place where in fact it is being generated in another which may be quite distant from the place where the sound is loudest. Broadly the causes of noise are transmitted pump noise, worn out bearings, pipe creep, air in the pipes and resonance. Of these resonance is the loudest and often the most bizarre.

Pump noise

Pump noise is a low humming or grumbling noise that can be heard continuously while the central heating circulating pump is on and stops when it is turned off. It may be loudest near the pump but equally may be transmitted along the pipes connected to it and then be most audible at a point where the pipe is in contact with a panel that acts as a sounding board.

Pipe creep

Pipe creep is caused by the changing temperature of pipes that occurs as hot and cold water moves round them. In response to this they expand and contract and this makes the movie in and out of the gaps through which they pass. This leads to a creaking noise which is most noticeable after the system has been turned on or off. Most of the time this noise is easily recognisable and is completely benign. Nothing more than reassurance is necessary. If there is a strong desire to correct the problem it can be extremely difficult to track it down. It usually amounts to trial and error or trying to listen to where the noise is loudest and identify which part is responsible by placing rubber between the pipes and beams to see if the noises muffled or finding which part is changing temperature rapidly at the time when the noise is most troublesome. It may be found that the noise is occurring where a pipe passes through a beam of wood. If this is the case the joint can be lubricated with a dry lubricant such as graphite or even talc which will usually solve the problem.

Resonance

Resonance causes a wide range of noises from squeaking and whistling from radiator valves to low frequency knocking that can be heard all over the house. Resonance is a feature of central heating systems with a relatively high pressure difference between the flow and returned sides. This is particularly prone to occur in modern systems that use thermostatic radiator valves. System installed with radiators only using thermostatic valves are virtually guaranteed to suffer some form of resonant noise which is why it should not be done. The problem arises because the radiator valves turn on and off in response to the temperature in the room and also in response to the pressure across them. When the central heating system is on and the house warm the radiator valves will all turn off. This will leave no open pathway between the flow and return sides. As explained in chapter 5 this leads to increased differential pressure. Thermostatic Radiator valves are usually operating on the margin between open and closed. As a valve closes it goes through a phase of being nearly closed then suddenly completely closed. At the moment of complete closure a brief high-pressure is created just behind the valves as the column of water moving towards it is stopped. This high-pressure zone is reflected back towards the boiler. When the wave reaches the boiler or expansion chamber it is reflected back again towards the radiator. When it reaches the radiator it can open the valve momentarily but as it is a pressure wave not a static pressure the effect is transient and the valve immediately closes and the cycle is repeated leading to knocking, the frequency of which depends on the distance along which the wave travels to and from the expansion chamber or whatever other point it is being reflected from. Knocking frequencies are usually between 1 and 10 Hz. The resonant circuit is analogous to a very long water filled clarinet where the radiator valve acts as the reed.

Knocking is a feature of thermostatic radiator valves but not a symptom of their malfunction. When knocking starts a radiator valve can be found which seems to be causing it because the knocking stops when the valve is turned up or down. Do not be tempted to replace it as the problem is with the system not that particular valve and all the others will be equally affected under slightly different circumstances.

The solution is to reduce the pressure differential between the flow and return sides. There are two ways are doing this. Perhaps the most obvious is to fit a differential pressure release valve. This works very well and not only stops knocking but also reduces a lot of the hissing and whistling associate with thermostatic radiator valves generally. It has the disadvantage that it cannot, or should not be used with a condensing boiler. Condensing boilers rely on a relatively cool return temperature as the water from the return side is used to condense water vapour from the flue gases. If this temperature is to high much of the water vapour will not be condensed and its latent heat lost, reducing boiler efficiency. Differential pressure release valves divert hot water from the flow to the return side without passing through any radiators and so without cooling it significantly. Consequently with high bypass flow rates they significantly raise the temperature of the return to the boiler resulting in loss of efficiency.

The alternative method to use if a condensing boiler is installed is to remove the thermostatic radiator valve from one of the radiators and replace it with a lock shield valve or manual valve that is always open. The problem with this is that the radiator is then permanently on with no thermostatic function so it must be in a place where there is a thermostat that controls the whole system such as a hall or living room.

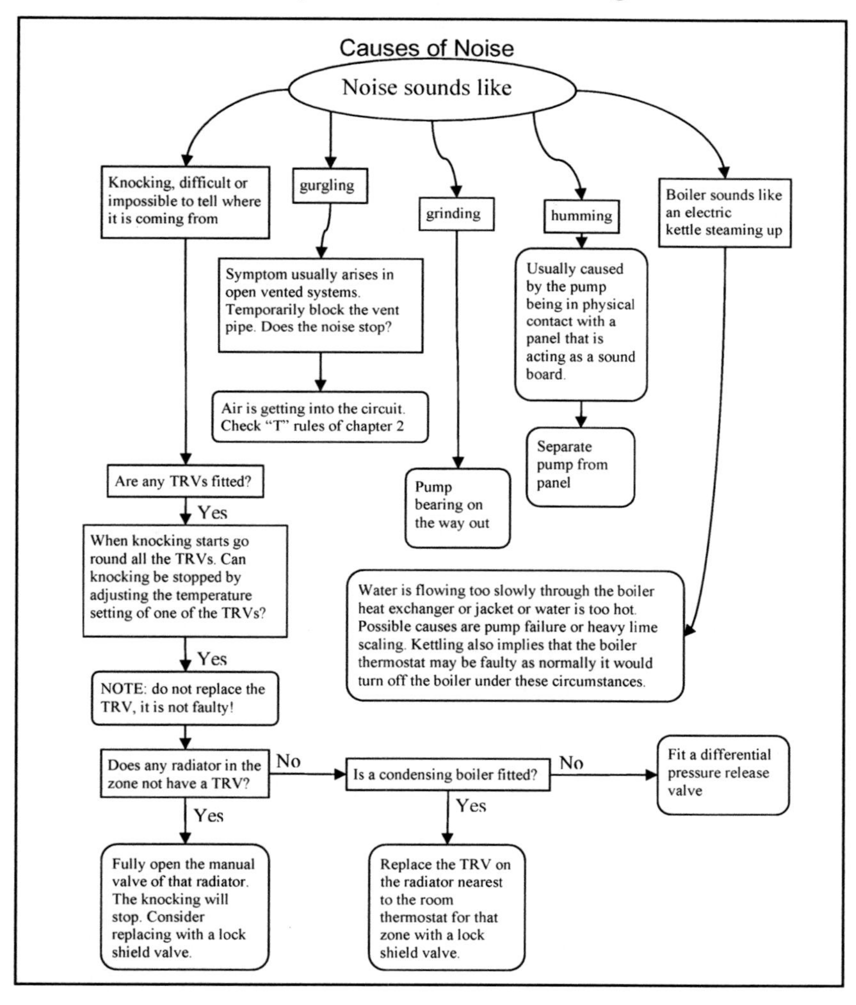

Reverse circulation fault diagnosis

Reserve circulation faults cause radiators to heat up unbidden when the central heating is off but the hot water system is on. They are a consequence of poor central heating installation rather than component failure. Reverse circulation occurs when convective flow drives heated water through part of the space heating circuit. It is caused by installing the return connection to the radiator system between the boiler and the hot water tank connections. When the space heating is turned off the control valve is closed so no hot water will enter the flow side of the radiator circuit. Suppose that the return side of the radiator circuit is plumbed into the return to the boiler between the boiler and the return from hot water cylinder. Although the primary hot water return pipe is on the cool side of the primary hot water circuit, the water flowing in it will be quite hot especially when the cylinder temperature is reaching fully heated. This hot water will then flow past the open pipe connected to the return side of the radiators. In the radiator circuit the water is cold and hot water, being lighter, will flow up the return pipe into the radiator circuit. What happens then will depend on whether the radiators are individually turned on. Usually they are as space heating is more conveniently controlled centrally than by turning radiators on and off individually. In that event the hot water will flow into radiators from the return end and flow out from the flow end into the flow side of the radiator circuit. As the inlet of the flow side of the radiator circuit is closed the only place the hot water can go is back into another radiator; this time in the normal direction. It ends cooler than it was, back in the return side of the radiator circuit. Radiator circuits are complex because they have to ramify all over the house so it is quite likely that a second returned pipe will be available to carry the water back down to level of the boiler.

To find out if the fault is caused by reverse circulation first see if the relevant motorised valve is working properly. If it is there are some things about reverse circulation that may help with diagnosis. Firstly at least one radiator heats up in the wrong direction i.e. from the return end to the flow end rather normal from flow to return. If a radiator that is doing this can be identified the reverse flow can be confirmed by provoking reverse flow, then when the radiator is warms turn on the central heating at which point the flow side of the radiator should warm up faster to become hotter than the other side as hot water moves from the flow to return site in the normal way reversing the previous convective heating effect. This test can be more difficult in practice that it sounds in theory because some of the radiators may heat in the correct direction and even those that heat in the wrong direction warm up more slowly than they do when operating normally making it difficult to tell which end is hottest.

Reverse flow is avoided if the hot water flowing to and from the hot water tank does not pass the radiator connections. This is the origin of the sequences of the 3T and 4T2T rules explained in chapter 2.

Index

16th century, 5
18th century, 7
18th-century, 5
19th century, 7, 8
20th century, 8, 9
21st century, 7, 8
22nd century, 8
23rd century, 7, 8
350 BC, 3
3T layout, 29, 30, 31, 32, 181
4T 2T layout, 31, 32, 33, 34, 181

A

acceleration due to gravity, 15, 18
accelerator, 138, 139
air in drawing, 17, 29, 30, 31, 32
algae, 53, 80
aluminium, 76, 77, 80
anode, 75, 76
anodic, 75, 83
anticipator. *See* accelerator, 139
aqueous solution, 73
atom, 74

B

bacteria, 53, 80
bar (uit of pressure), 15, 78, 106, 115, 116, 124
bath, 9, 41, 47, 56, 125
biofilm, 54, 134
boiler, 128-137, 163-165
  high resistance, 22, 30, 31, 32, 123
  internal thermostat, 124, 127, 128, 148
  low resistance, 17, 18, 22, 30, 63, 123
boiler interlock, 10, 127, 149, 150, 151, 153
Bolton, Matthew, 5
British Isles, 5, 7
British Thermal Unit, 98
BTU. *See* British Thermal Unit, 46, 98
Bunsen, Baron, 8

C

capillary fittings, 89, 90
carbon monoxide, 7, 55, 123
cast iron, 6, 17, 18, 20, 22, 30, 32, 62, 63, 64, 115, 123, 126, 139
cathode, 75, 76
cathodic, 75, 83
centigrade, 98
chloride, 73, 77
chlorination, 80

circuit, 37-52
S plan, 25, 153, 154, 155, 168
single pipe, 5, 13, 18, 20, 21, 22, 23, 27, 81
two pipe, 13, 20, 23, 31, 106
W plan, 24, 25
Y plan, 24, 25, 152, 168
cistern, 9, 14, 16, 17, 19, 27, 28, 32, 33, 35, 38, 40, 41, 42, 47, 49, 68, 73, 79, 82, 83, 124, 129, 130, 131, 132, 133, 134, 135
citric acid, 79
clean burning, 7
coal, 6, 7, 8, 9, 103, 120
coal gas, 7, 8, 9
coefficient of thermal expansion, 20
cold feed, 14, 16, 17, 19, 28, 29, 30, 31, 32, 33, 40, 42, 68, 79, 81, 115, 134
cold water, 14, 18, 28, 33, 40, 45, 47, 48, 50, 51, 54, 56, 57, 67, 68, 78, 83, 125, 129, 130, 131, 132, 134, 148, 177
Colonel Coke, 5
combi boiler, 36, 45, 46, 47, 48, 56, 124, 125, 126, 129
combination boiler. *See* combi boiler, 9, 45, 46, 114, 125, 126
comfortable temperature ranges, 51
condensing boiler, 59, 65, 77, 100, 102, 120, 121, 125, 178, 179
control circuit, 25, 124, 140, 148, 150, 151
convection, 5, 9, 13, 15, 23
conventional boiler, 33, 45, 46, 120, 121
copper, 9, 26, 37, 38, 39, 40, 51, 56, 57, 67, 73, 74, 75, 76, 78, 85, 86, 88, 89, 90, 91, 92, 93, 94, 95, 168
copper pipe, 26, 40, 56, 57, 67, 85, 86, 88, 89, 90, 91, 93, 95
corrosion, 17, 21, 30, 35, 41, 68, 69, 73, 76, 77, 78, 80, 81, 83, 84, 89, 110, 129, 131, 132, 134
cylinder thermostat. *See* tank stat, 52, 109

D

datum, See pressure datum, 16, 17, 19, 28
deburring tool, 90
degrees, 97, 98
density, 7, 15, 17, 18, 20, 40, 47, 99, 159
descaler, 79
Desdga, Peter, 8
dezincification, 76
differential pressure release valve, 26, 65, 178
direct cylinder, 43, 44
diverter valve, 24, 25, 105, 106, 109, 146, 151
draft, 124
dry fire, 28, 127
dust, 5

E

earthing, 56, 57
efficiency, 1, 2, 9, 10, 11, 39, 47, 69, 84, 102, 103, 119, 121, 125, 138, 156, 158, 164, 178, 179
electric shower, 161
electrical safety, 145
electrochemical cell, 73, 74
electrodes, 73, 74, 75, 76
electrolyte, 73, 74, 75
electronic lime scale remover, 79
expansion chamber, 34, 124, 137, 178

F

fireplace, 102, 103
flame failure device, 164, 165
floor screeding, 71
flue, 120, 123, 124, 126, 127, 159, 165, 178
  balanced, 122, 124, 126
  room sealed, 122, 123
  room vented, 123
frost protection, 154, 155
fuel, 3, 7, 9, 10, 39, 55, 56, 100, 103, 119, 120, 121, 123, 127, 142, 157
fully pumped, 10, 13, 15, 24, 27, 39, 109
fully pumped (system), 10, 13, 15, 24, 27, 39, 109
fungi, 80

G

galvanic corrosion, 67
Galvanic corrosion, 67, 73, 75, 76
gas, 2, 3, 7, 8, 9, 10, 27, 30, 32, 34, 40, 42, 43, 45, 55, 56, 64, 65, 67, 68, 76, 85, 100, 102, 113, 119, 120, 121, 122, 123, 126, 127, 139, 158, 159, 163, 164
Gas Light and Coke Co, 7
gas lighting, 7, 8, 9, 122
global warming, 2, 11
grate, 103
  wood burning, 103
gravity, 6, 10, 15, 23, 37, 38, 39, 43, 47, 80, 81, 95, 98, 121, 126
gravity cleaning, 80, 81
gravity fed, 6, 10, 37, 38, 39, 43, 47
gravity fed hot water, 37

H

haemoglobin, 55
hard water, 44, 78
heat, 97
heat capacity, 5, 6, 21, 24, 26, 47, 98, 99, 100, 115, 139, 159
heat exchanger, 9, 13, 24, 28, 32, 38, 39, 43, 45, 46, 47, 48, 49, 54, 68, 77, 78, 114, 120, 122, 125, 127, 128, 141, 148, 159, 165
high-pressure steam, 5
history, 3
hot water, 37
hot water cylinder, 29, 38, 48, 51, 52, 56, 80, 85, 125, 129, 138, 140, 148, 157, 172, 180
hot water priority, 25, 37, 106, 125, 150
hot water system, 9, 13, 24, 25, 27, 37, 38, 39, 43, 45, 46, 49, 50, 51, 52, 54, 56, 78, 79, 81, 106, 124, 126, 132, 155, 160, 180
hot/cold interface, 39
Housteads Roman Fort, 4
hydrodynamics, 116
hydrogen, 7, 54, 67, 74-76, 100, 120
hydrogen electrode, 74
hydrostatic pressure, 16, 18, 19, 40, 67, 113
Hydrostatic pressure, 15
hydrostatics, 15, 116
Hypocaust, 3, 4

I

immersion heater, 24, 35, 43, 44, 48, 49, 52, 125, 159
inhibitor, 67, 68, 73, 78, 80, 83, 84, 94, 117, 134

insects, 40, 41, 54, 131, 132
insulation, 10, 33, 35, 48, 51, 52, 56, 70, 130, 165
ion, 74
iron, 62, 73, 74, 75, 76, 78, 81, 85, 95

J

Joule, 5, 97, 98

K

Kell, J. R., 18, 23
Kelvin, 97, 98
key to control circuits, 119, 169
kilowatt hour, 97
kitchen sink, 115, 132
knee pads, 58
Kyoto agreement, 11

L

latent heat, 93, 100, 120, 178
Legionella, 33, 48, 51, 53, 54, 80, 129, 131, 135
legislation, 11, 55, 120
lime scale, 30, 47, 48, 49, 50, 51, 52, 68, 78, 79, 81, 84, 114
lock shield valve, 63, 65, 66, 68, 81, 179

M

mains pressure hot water, 9
masonry heater, 3, 4, 121
methane, 7, 9
micro bore, 26
micro CHP. *See* micro combined heat and power, 2, 10
micro combined heat and power, 10
micro leaks, 68
mid position valve, 24, 25, 105, 109, 114, 146, 147, 152
mini-programmer, 142, 148, 150, 151
motorised valve, 2, 24, 37, 78, 105, 106, 109, 110, 125, 137, 140, 143, 168, 172, 180
  Honeywell V4073, 105
  manual lever, 107, 108
  plumbing part, 106, 107, 110
Murdoch, William, 7

N

natural gas, 7, 9
negative pressure, 30
Newton (unit of force), 15
noise, 9, 22, 26, 55, 65, 111, 113, 114, 130, 165, 177, 178
Northumberland, 4, 62

O

off peak electricity, 43, 44, 158, 160
oil, 2, 7, 8, 43, 54, 56, 85, 95, 102, 119, 120, 121, 122, 124, 126, 158, 159
open vented, 16, 19, 24, 27, 28, 30, 31, 32, 33, 42, 68, 81, 82, 83, 133
overflow, 130, 131, 132, 133, 134
oxidation, 33, 73, 94
oxygen, 17, 33, 35, 55, 73, 78, 129

P

Pa. *See* Pascal, 15, 18, 20
partly pumped, 10, 13, 23, 27, 32, 148, 150, 151
Pascal (unit of pressure), 15
pH, 54, 77, 120

Philadelphia, Pennsylvania, 53
pilot light, 164, 165
pipe cutter, 88, 90, 95
Platt, Hugh, 5
pollution, 7
power flushing, 80, 81, 82
pre-mixed gas burner, 8
pressure, 17-29, 112
  atmospheric, 8, 15, 67, 101, 113, 115, 158
pressure datum, 16, 17, 19, 28
primary hot water circuit, 13, 23, 24, 25, 28, 29, 32, 33, 38, 43, 49, 54, 82, 83, 105, 106, 127, 148, 151, 153, 159, 180
primatic cylinder, 81
pump, 120
  three speed, 111, 112, 114, 115
pump-over, 17, 28, 29, 30, 31, 32, 33, 68

R

radiator, 59-68
  cast iron, 62, 63, 115, 123
  heat output, 59, 60, 61, 63, 66, 101, 102, 103, 120, 160
  paint, 59
  pressed steel, 59, 60, 61, 63
  roll top, 61, 62
  venting, 33, 67
radiator valve, 17, 63, 65, 66, 81, 82, 111, 177, 178
radiators
  balancing, 22, 41, 65, 66, 67, 111
RCD (Residual current device), 57, 58, 161, 168
renewable fuel, 3
response time, 5, 114
reverse circulation, 29, 180
reverse flow, 32, 180
RMS (rood mean squared), 55
Roman Hypocaust, 3

S

sacrificial electrode, 77
scale, 2, 9, 77, 78, 79, 95, 98
sealed, 5, 16, 17, 27, 34, 35, 42, 78, 116, 123, 124, 129
SEDBUK, 10
Sharp, James, 8
shower, 161
  balancing, 22, 41, 65, 66, 67, 111
SI units, 98
single feed cylinder. *See* primatic cylinder, 42
siphoning, 40, 130, 131
sludge, 76, 77, 81, 84, 113, 114, 117
soapstone, 99
solder ring fittings, 89, 90, 91
solid fuel, 1, 3, 6, 7, 8, 102, 119, 121, 124
space heating, 9, 11, 13, 25, 27, 28, 29, 30, 32, 33, 37, 39, 42, 48, 82, 85, 105, 106, 109, 114, 115, 124, 125, 127, 142, 148, 149, 151, 152, 153, 155, 157, 159, 172, 180
specific heat capacity, 98, 99
stainless steel, 56, 77, 81
static pressure, 17, 27, 114, 115, 178
steam based systems, 5
storage cistern, 42, 45, 54, 131, 134
storage heater, 3, 159, 160
  open vented, 16, 19, 24, 27, 28, 30, 31, 32, 33, 42, 68, 81, 82, 83, 133
  single pipe, 5, 13, 18, 20, 21, 22, 23, 27, 81
system boiler, 35, 36, 119, 124

T

tank stat, See thermostat, hot water cylinder, 51
TDS (total disolved solids) meter, 80, 82, 83
temperature, 55
temple of Ephesus, 3
thermal efficiency, 3
thermal expansion, 69, 133
thermal store, 9, 36, 37, 46, 47, 48, 49, 50, 81, 125, 126, 129, 159
thermal store hot water, 9, 36, 37, 46, 47, 48, 49, 50, 81, 125, 126, 129, 159
thermostat, 137-143
  frost, 129, 141, 142, 145, 153, 154, 155
  hot water cylinder, 51
  low limit, 141, 142, 145, 153, 155
  room, 138, 140, 141, 145, 149, 150, 151, 152, 153, 155, 168, 172
thermostatic radiator valve, 64, 65, 72, 78, 81, 148, 178, 179
three port valve, 24, 105, 109, 152, 153
total dissolved solids meter. *See* TDS meter, 80, 84
TRV, See thermostatic radiator valve, 64, 65
tube bending machine, 86, 87
twin entry radiator valve, 81
two jar test, 67
two pipe, 13, 20, 23, 31, 106
two port valve, 25, 105, 109, 146, 155

U

U tube, 40
under-floor heating, 69, 70, 71
un-pumped, 13, 18, 21, 22, 23
un-vented cylinder, 9, 50, 129
upgrades, 1

V

vent, 14, 17, 27, 28, 29, 30, 31, 32, 33, 34, 35, 40, 41, 45, 81, 83, 115, 116, 117, 130, 131, 132, 133, 134
vent pipe, 17, 27, 28, 29, 30, 40, 45, 132, 133, 134
vented, 9, 17, 24, 27, 28, 33, 37, 39, 40, 41, 45, 46, 50, 67, 73, 117, 122, 123
ventilation, 55, 101, 123
ventilatory heat loss, 101, 102, 103
ventricular fibrillation, 55, 57
volumetric heat capacity, 47, 98, 101

W

water jacketed heater. *See* thermal store, 46
water pressure, 16, 26, 28, 45, 46, 129
water softener, 77, 78, 80
water vapour, 10, 100, 120, 121, 178
Watt (unit of power), 5, 97
wire brush, 89

Y

yeasts, 80

Z

zinc, 76
zoning, 10, 25

LaVergne, TN USA
18 June 2010
186646LV00004B/113/P

9 781904 623625